GLOBALIZATION, INSTITUTIONS, AND REGIONAL DEVELOPMENT IN EUROPE

The **European Science Foundation** is an association of its fifty-nine-member research councils, academies and institutions devoted to basic scientific research in twenty-one countries. The ESF brings European scientists together to work on topics of common concern, to coordinate the use of expensive facilities, and to discover and define new endeavours that will benefit from a cooperative approach.

The scientific work sponsored by ESF includes basic research in the natural sciences, the medical and biosciences, the humanities and the social sciences.

The ESF links scholarship and research supported by its members and adds value by cooperation across national frontiers. Through its function as a coordinator, and also by holding workshops and conferences and by enabling researchers to visit and study in laboratories throughout Europe, the ESF works for the advancement of European science.

This volume arises from the work of the ESF Programme on Regional and Urban Restructuring in Europe (RURE).

Further information on ESF activities can be obtained from:
European Science Foundation
1 quai Lezay-Marnésia
67080 Strasbourg Cedex
France
Tel. 88 76 71 00
Fax 88 38 05 32

GLOBALIZATION, INSTITUTIONS, AND REGIONAL DEVELOPMENT IN EUROPE

Edited by

ASH AMIN

and

NIGEL THRIFT

OXFORD
UNIVERSITY PRESS

10 -31 -6

This book has been printed digitally and produced in a standard design in order to ensure its continuing availability

OXFORD
UNIVERSITY PRESS

Great Clarendon Street, Oxford OX2 6DP

Oxford University Press is a department of the University of Oxford.
It furthers the University's objective of excellence in research, scholarship,
and education by publishing worldwide in

Oxford New York

Athens Auckland Bangkok Bogotá Buenos Aires Cape Town
Chennai Dar es Salaam Delhi Florence Hong Kong Istanbul Karachi
Kolkata Kuala Lumpur Madrid Melbourne Mexico City Mumbai Nairobi
Paris São Paulo Shanghai Singapore Taipei Tokyo Toronto Warsaw

with associated companies in Berlin Ibadan

Oxford is a registered trade mark of Oxford University Press
in the UK and in certain other countries

Published in the United States
by Oxford University Press Inc., New York

Preface and Introduction

ONE widely circulated story is that local and national identities and autonomies are becoming increasingly shaped by and sacrificed to global forces. These forces might include transnational corporations, international communications networks, global flows of ideas and images, global financial institutions, and influential international regulatory authorities such as the IMF, G7, or the European Commission. Another, and contrasting, story is that the local economy and local institutions are becoming more, rather than less, salient. Here, it is argued that the hitherto dominant national economy and nation state is losing its organizational relevance under the weight of local socio-political movements, the reinforcement of local identities in the face of global 'intrusions', and the resurgence of self-contained regional economies such as the Italian industrial districts or Silicon Valley in California. Thus, one story tells of the salience of global space, whilst the other tells of the salience of local space.

The purpose of this book is to interrogate these two stories through the specific example of the implications of globalization for local economic prospects. Our purpose is to move beyond these polarized accounts by stressing the institutional dimensions of current changes. In other words, our intention is to develop an institutionalist focus on the problem and prospects of European cities and regions in a global political economy, as a way of exploring the interrelationship between globalizing and localizing tendencies.

The book therefore draws upon current debates and developments in economics, sociology, management studies, politics, and geography that argue that economic life is 'embedded' in social relations and is therefore heavily dependent upon a mix of cognitive, cultural, social, and political institutions.

In particular the papers in this book develop the idea that the performance of local economies in a globalizing world is critically dependent upon their 'institutional thickness': this term has a variety of meanings which are more fully articulated in Chapter 1. The question explored in this book is what this 'thickness' amounts to and how it affects local economic development paths. Accordingly some chapters draw on the experience of successful local areas in Europe, to identify the sorts of institutional conditions which facilitate their survival in the wider economy. Other chapters, in contrast, problematize the role of local institutions, to illustrate the apparent inability of established traditions and practices, especially in some less prosperous European areas, to facilitate an improvement in the terms of engagement with the global economy. Thus, in recognizing the role of institutions in local economic life, the book carefully seeks to avoid either a functionalist or an over-parochial interpretation of their significance.

The first three chapters of the book provide a more general introduction to the problems of theorizing local economic development in a globalizing world.

In Chapter 1, Ash Amin and Nigel Thrift provide a framework for interpreting local economic trajectories and development possibilities within the context of an all-pervasive globalization of economic, political, social, and cultural processes. The authors stress that globalization does not represent the end of the local. They stress the possibility of local 'proactivity' constructed around strong structures of social and economic representation, but also highlight that only particular forms of 'institutional thickness', which are not universally available, are conducive for growth drawing upon local resources. The chapter, thus, seeks to problematize the role of 'thickness' in local economic development.

Chapter 2, by Peter Dicken, Mats Forsgren, and Anders Malmberg, examines the changing role of perhaps the most powerful institution shaping economic life today, namely the transnational corporation, as an agent of local economic development. It reviews critically the new literature which claims that the most innovative corporations now display high levels of 'embeddedness' within their host regions, and goes on to argue that often local commitment falls far short of providing any real stimulus for self-sustaining growth.

Chapter 3, by Sarah Whatmore, goes on to consider the contradictions in a very different context associated with the integration of local areas into global economic networks. It examines the emerging tension between production-based and consumption-based (e.g. tourism) uses of the countryside, in the context of the refashioning of rural areas within global agro-commodity chains dominated and manipulated by large corporations. Her analysis stresses the contested nature of global change, and the consequent openness of local possibilities.

The three chapters which follow examine, in different ways, the assumption that future industrial competitiveness will rely on vertically disintegrated and localized industrial networks, supported by local institutions. This question is approached through an examination of the processes of restructuring affecting some of the most celebrated examples of self-sustaining regional economic development in the course of the last two decades. Chapter 4, by Franz Tödtling, tackles the question of whether the so-called innovation-intensive post-Fordist economy necessarily relies on highly localized entrepreneurial and institutional arrangements. It comes to the conclusion, on the basis of a review of the experience of the most notable examples of local success (technopoles, industrial districts, large-firm agglomerations), that the geography of innovation is both global and local, and selective of areas with strong entrepreneurial traditions and a local milieu possessing a robust set of institutions which encourage education, training, firm-formation, and technology transfer.

The anatomy and future of the areas of local economic success which have displayed sophisticated levels of local institutional formation in recent years is discussed in more detail in Chapters 5 and 6. Chapter 5, by Philip Cooke and Kevin Morgan, focuses on the regions of Baden Würtemmberg and Emilia-Romagna to flesh out the main institutional characteristics of the so-called 'intelligent' region, which is capable of securing high levels of innovation, economic adaptability, and local embeddedness of industry. Their account, however, is cautious since it highlights new pressures on such regions forcing the globalization of industry and posing a major challenge to the hitherto dominant role of close local networking among and between firms and other institutions.

Similarly, in Chapter 6, Amy Glasmeier discusses the implications of transformations in the international economic and technological regime surrounding an industry, upon the capability of local institutions which were once the bedrock of local economic success. The chapter exposes the obstacles posed by long-standing craft-based traditions and disintegrated production systems within the Swiss watch industry, acting to prevent a paradigmatic shift towards a global competition based on microelectronics and scale economies. It is a chapter which challenges a growing consensus that accepts, as an indisputable truth, the economic strength of craft-based industrial districts.

The theme of local institutional specificity and its influence on the economic success of an area is further elaborated in Chapter 7, which examines the dynamics of local labour-market regulation and reproduction. Jamie Peck evaluates the widely accepted view that the non-mass-production economy requires flexible labour-markets and investment in skills and training, to argue that these are conditions which are never automatically guaranteed, but the product of institutionally embedded practices and codes of conduct at both national and local level. Such 'rooted-ness' for Peck is what differentiates the character and economic efficiency of local labour-markets. This emphasis on the historicity and spatial specificity of institutions underpinning labour-markets puts into question the sagacity of the current policy belief that an 'injection' of new skills and training into local labour-markets will bring automatic economic rewards.

Chapters 8 and 9, in contrast to preceding ones dealing with the areas of economic success, explore the problems of economic revitalization in areas which have long lost their industrial 'coherence' and institutional 'cohesive-ness': attributes which can be drawn upon to develop local economic 'integrity' in the global political economy. Gernot Grabher, in Chapter 8, outlines the paltry local economic benefits and spin-off associated with the 'truncated' industrialization of the East German regions in the hand of

western investors; a process of inward investment which has swept aside or
bypassed old socio-institutional traditions, which once, at the very least,
guaranteed a minimum level of local industrial 'embeddedness'.

Chapter 9, by Ray Hudson, builds on the theme of local institutional
impoverishment and loss of economic cohesion, by examining the
fractures within areas which have for long had to play to the tune of many
successive global forces. It evaluates, within the context of north-east
England, the experimentation over the last decade with a plethora of
initiatives ranging from small firm support, policies to attract 'flagship'
transnational companies, recreation and leisure-based projects, and
initiatives aimed at improving the quality of life of middle-class residents.
With its emphasis on local economic fragmentation and the narrow
sectarianism of the institutions which dominate the region, the chapter
takes a pessimistic stance on the prospects of success of local institutional
solutions.

The penultimate chapters of the book transcend narrow definitions of
local development, most especially through their emphasis on culture and
on politics in the formation and institutionalization of local identities. In
Chapter 10, Kevin Robins and James Cornford explore the implications of
globalization upon the production of cultural goods as well as cultural
identity, at the local level. Focusing on the audio-visual industries, they
illustrate how initiatives at the local level are very much at the thin end of a
wider process of global production and consumption of cultural goods. A
central concern of this chapter is to show how new dynamics within the
audio-visual industries are serving to construct complex, and often
fragmented, local cultural identities.

In the eleventh chapter, Costis Hadjimichalis draws on examples from
the Mediterranean to tell the story of regional social and political
movements against what are perceived to be 'global intrusions'. The
chapter makes the important point that social mobilization is an essential
condition for the construction of a local civil society which is capable of
shaping and managing its own affairs, or, at the very least, engaging with
some authority in the global political economy.

The final chapter, by Ash Amin and Nigel Thrift, reflects on some of the
major themes associated with the book. Most particularly it takes up the
issues of institutional diversity, institutional effectiveness, and institutional
embeddness as determinants of local economic success.

Acknowledgements

THIS book has been produced under the auspices of the European Science Foundation's Regional and Urban Restructuring in Europe (RURE) Programme. We would like to thank Anders Malmberg and Peter Dicken for their help with the production of this book. Above all we are grateful to the authors of each chapter in this book, who have borne with us through numerous discussions, and revisions of drafts.

A. A. and N. T.

Newcastle upon Tyne

Contents

Notes on Contributors

DR ASH AMIN is Senior Lecturer in Geography at the University of Newcastle upon Tyne.

PHILIP COOKE is Professor of Regional Development and Director of the Centre for Advanced Studies at the University of Wales, Cardiff.

JAMES CORNFORD is Research Associate in the Centre for Urban and Regional Development Studies at the University of Newcastle upon Tyne.

PETER DICKEN is Professor of Geography at Manchester University.

DR MATS FORSGREN is Lecturer in Economics at Uppsala University.

DR AMY GLASMEIER is Associate Professor in the Department of Geography at Pennsylvania State University.

DR GERNOT GRABHER is Senior Research Fellow at the Wissenschaftszentrum in Berlin.

COSTIS HADJIMICHALIS is Professor of Urban and Regional Planning at the University of Thessaloniki.

RAY HUDSON is Professor of Geography at Durham University.

DR ANDERS MALMBERG is Research Fellow in the Department of Geography at Uppsala University.

DR KEVIN MORGAN is Senior Lecturer in the Department of City and Regional Planning at the University of Wales, Cardiff.

DR JAMIE PECK is Lecturer in Geography at the University of Manchester.

KEVIN ROBINS is Lecturer in Geography at the University of Newcastle upon Tyne.

NIGEL THRIFT is Professor of Geography at Bristol University.

DR FRANZ TÖDTLING is Lecturer in the Department of Spatial Organization, University of Economics, Vienna.

DR SARAH WHATMORE is Reader in Geography at Bristol University.

List of Figures

List of Tables

Map

1

Living in the Global

Ash Amin and Nigel Thrift

INTRODUCTION

'Globalization' is now seen too often as an all-pervasive force in the modern world. Yet, even now, there is still considerable dispute over what the term might mean, let alone what it might portend (Robertson 1990; King 1991). Commentators seem willing to agree that the term must be linked in some way to the wilting of the idea of a cohesive and sequestered national economy and society (Giddens 1989). They seem willing to agree that a global economy and society now exists, and not just in embryonic form. They seem convinced that as a result of these changes, everyday life, in developed countries at least, has, in some senses, changed, if only because no one can opt out of the changes taking place (Giddens 1991). Finally, commentators seem sure that there is a 'dialectic' between the global and the local, that in some sense what counts as the local has been transformed by globalization (Massey 1991).

But these are all emblematic (and enigmatic) indices of global change. They do not constitute a finished theory of globalization; rather, they point to a set of concerns which a number of commentators have in common. In one form or another, these concerns, now increasingly voiced across the social science disciplines, refer to the real prospects for self-governed development paths in an era of global interdependencies. Thus, for example, sociologists and anthropologists increasingly have been concerned with the implications of international mass migrations or the use of global media corporations for local cultural identities and identity formation. Similarly, industrial economists, industrial geographers, and business analysts have been keen to explore the significance of the spread of transnational corporations and the rise of new global business strategies on smaller enterprises and local industrial agglomerations. In politics, too, increasingly, discussions on regional or national institutions are centred around whether the rise of supranational political bodies threatens the space of local institutions and local political struggles. Thus, while some observers perceive globalization in its various manifestations as a threat to local diversity and local autonomy, others argue that it rather represents a change in the context in which local development paths are articulated.

The aim of this chapter is to bring together these diverse concerns and to begin to develop a conceptual framework which helps in making sense of the economic development prospects of regions in an era of pervasive globalization. More specifically, our aim is to explore the ways in which the global context of cities and regions is changing and to discuss what this change might represent in terms of the development prospects of different types of local economy. Thus the chapter begins with an account of the principal ways in which economy and society can be said to have become 'globalized'. This is followed by a discussion of the degree to which this process of global integration represents a retention of diversity and difference at the local level. We argue that globalization does not represent the end of territorial distinctions and distinctiveness, but an added set of influences on local economic identities and development capabilities.

The third section, drawing upon the example of areas which continue to remain power-houses and centres of agglomeration within the globalized economy, seeks to identify the factors which allow certain places to retain a high level of local economic 'integrity'. In explaining the possibility of place-centredness, that is local agglomeration within global production filieres, particular emphasis is given to the role of certain institutional conditions, ranging from strong local institutional presence through to the strength of shared rules, conventions, and knowledge. The final section goes on, however, to suggest that the degree of 'institutional thickness' necessary for places to become locally embedded in the context of global economic structures is not available or reproducible ubiquitously. This raises a difficult policy dilemma, that the vast majority of cities and regions will continue to occupy specific functional positions within a global division of labour—a status which is not likely to promise self-reproducing growth at the local level.

The chapter, thus, seeks to bring together different literatures on globalization and confront them with questions concerning the future of local economies. As such, it also constitutes a conceptual framework for the remaining chapters of the book which deal with the implications of different aspects of globalization on the development prospects of different types of regional economy in Europe.

THE GLOBAL ECONOMY

Most writers seem to agree that the transition from an international to a global economy can be dated from the early 1970s and the breakup of the Bretton Woods system of control of national economies. We can now see some of the fruits of the subsequent reorganization of the world economy. There are, of course, many of these but seven aspects can perhaps be counted as the most important, in terms of both geography and pervasiveness of influence.

The first of these is the increasing centrality of the financial structure through which credit money is created, allocated, and put to use, and the resulting increase in the power of finance over production. Thus Harvey (1989) writes of the degree to which financial capital has become an independent force in the modern world, while Strange (1991) writes of the increased 'structural' power exercised by whoever or whatever determines the financial structure, especially the relations between creditors and debtors, savers and investors. It is particularly the global reach of finance which is striking today, as global money, in a variety of guises, straddles across and regulates the world's national economies and business transactions.

The second aspect is the increasing importance of the 'knowledge structure' (Strange 1988) or 'expert systems' (Giddens 1990). Everywhere in economics, sociology, economic sociology, and so on, more attention is being paid to the importance of knowledge as a factor of production. This debate can be found in the form of arguments over whether an increase in the educational base of workforces produces higher GNP or in arguments over the importance of learning as a factor in efficient production, the communication of new technologies, the construction of conventions about how to produce, and so on. Whatever the form of the debate, it seems clear that the production and, later, distribution and exchange of knowledge is a crucial element of the global and local economic system on a scale that was never the case before. As with the financial structure, owing to the enormous interpenetration of know-how and scientific culture between nation states and the rise of enabling communications media and technologies, the 'knowledge structure' is becoming less and less tied to particular national or local business cultures, after it has been 'released' from its original context of formation.

A third and related aspect is the transnationalization of technology, coupled with an enormous increase in the rapidity of redundancy of given technologies. Especially in the knowledge-based industries such as telecommunications, aerospace, pharmaceuticals, and financial services, the development of micro-electronics applications has enormously quickened the diffusion of standards and know-how. At the same time, however, the increasing reliance on technological innovation resulting from the combination of new consumption norms, increased competition, and the systematization of the knowledge-base has forced the speed with which technological trajectories are required to change (Dosi, Pavitt, and Soete 1990; Metcalfe 1988). Thus at one level, the globalization of technology represents a levelling of access through the spread of electronic networks and information data bases. But, at another level, the greater uncertainty produced by this more complex environment militates against any except those institutions with considerable resources and continuous learning capacity. In other words, the new institutionalized risk environments (Giddens 1991) only reward the adaptable.

The fourth major global force is the rise of global oligopolies. There is a sense in much recent writing that we have now reached a point at which corporations have no choice but to 'go global' very early on in their careers, for at least three reasons (Strange 1991). First, because of new methods of production brought about by accelerated technological change. Second, because of the greater transnational mobility of capital which has made investing abroad easier, quicker, and cheaper. Third, because there have been major changes in the ease of transport and communication. The result is that national measures of concentration and market share have become less relevant as corporations manœuvre in global markets, with obvious consequences for the balance of economic power.

Fifth, and a parallel development to the globalization of production, knowledge, and finance, is the rise of transnational economic diplomacy and the globalization of state power. There is clearly a sense in which we have entered a new era in which governments and firms bargain with themselves and one another on the world stage. In addition, transnational, 'plural authority' structures like the UN, G7, and the EC have become increasingly powerful (Held 1991). The result appears to be the replacement of the hitherto prevalent system of international stabilization and rule formation based upon the unchallenged might of nations with the greatest 'structural' power (e.g. the United States), by a combination of issue-based agreements between members of 'plural authority' structures, and tentative gestures by the most dominant nations to occupy the centre stage in world political economy (Gilpin 1987). For some observers, this is a situation presaging great global institutional and regulatory uncertainty (Lash and Urry 1992; Robertson 1990).

Sixth, and related to the intensification of global communications and international migration, is the rise of global cultural flows and 'de-territorialized' signs, meanings, and identities. Appadurai (1990) talks of the rise and juxtaposition of global 'ethnoscapes' (the continual movement of tourists, immigrants, refugees, exiles, and guestworkers across the world); of global 'mediascapes' (the instantaneous worldwide distribution, by a diversity of electronic and non-electronic media, of one and the same pastiche of standardized commodities, generalized human values and interests, the cult of consumerism, and denationalized ethnic or folk motifs (see Smith 1990)); and of global 'ideoscapes' (key concepts and values from the Enlightenment world-view, such as 'freedom', 'democracy', 'sovereignty', 'rights', and 'citizenship', now all evoked by states and counter-movements in different local contexts to legitimate their existence and political aspirations). This is not a process of cultural homogenization in the hands of readily identifiable national 'ethnics' or ideologies, but the fusion of different master narratives (e.g. consumerism, cult of technology, Europeanism) and local vernaculars (e.g. separatism, folklorism, local sacred beliefs, etc.). These are cultural flows which have become 'deterritorialized', that is separated from their original local and national settings, but only to reappear in places as new influences blending in with existing local myths, memories, and beliefs (Smith 1990). Global culture, then, is a heterogeneous phenomenon, a juxtaposition of sameness and difference, of the 'real' and the 'imagined', of intersecting universal and local narratives (King 1991; Bauman 1992).

Finally, the result of the processes described above is the rise of new global geographies. There is an account, increasingly prevalent, in which the processes of globalization are seen to have produced borderless geographies with quite different breaks and boundaries from what went before. Whether we see the global economy as a 'space of flows' (Castells 1989), as almost without a border (Ohmae 1990), as a necklace of localized production districts strung out round the world (Storper 1991), as the centralization of economic power and control within a very small number of global cities (Sassen 1991), or as something in between these extremes, it is clear that geography is now globally local rather than vice versa.

At work, then, is a multi-faceted process of global integration guided by, but not always made in the image of, the most powerful transnational firms, institutions, actors, and cultural hegemonies of the capitalist world economy (Chesnaux 1992). The familiar imagery is the simultaneous production and consumption of the same products and images in every corner of the world (Clifford 1988), the restless flow of people, goods, finance, information, and ideas across the globe (Hannerz 1992), grafted on to metaphors such as 'global village' and 'one world' to convey messages of unity and common heritage.

THE LOCAL IN THE GLOBAL

A world of transnational flows, a world of global interconnections, a world of global-scale capitalist imperatives from which there is no escape. And yet, global economy and global society continue to be constructed in and through territorially bound communities, which, more importantly, continue to represent much more than the situations requirements of global forces. At a concrete level, metaphors such as 'global village' and 'one world' are complicated, perhaps even contradicted, by the presence of villages, towns, districts, cities, and regions which continue to tell their own stories of economic development and cultural or political distinctiveness. How, then, should we conceptualize the global–local nexus, that is the nature of the encounter between place and global space, and how should we think about the role of the individual locality in a globalized political economy? In what sense does 'territoriality' or place-boundedness matter? There is now a rich and diverse literature on what the local might mean in global terms, but this is not the place to review this literature. Instead, in what follows, we offer a series of observations about the 'local' designed to set the stage for the later discussion in this chapter on the scope for local economic 'integrity' in a global political economy.

Though in general we deploy the term 'local' to mean localities such as cities and regions rather than nations, and the term 'global' to signify worldwide processes, the terms are purposely left imprecise. This is because different phenomena will have different scales of representation. For the multinational corporation, for example, the local might represent the entire European market or a region in which a particular activity is located. As a signifier of cultural identity the local might represent a metropolitan borough with strong sense of local identification, or perhaps a region with distinctive socio-cultural traditions. As a player in the world economy, the local might include as small an area as the rural industrial districts of Italy or as large an agglomeration as Silicon Valley in California. Finally, the local as a unit of political representation might, in some contexts, refer to regions or cities with distinctive forms of political authority and governance structures or to areas on a smaller spatial scale which possess no formal political powers, but are inhabited by local élites which wield decisive political influence. Thus the ambiguity of definition of the 'local' or the 'global' is necessary and deliberate, precisely because, as Latour has emphasized, 'the words "local" and "global" offer points of view on networks that are by nature neither local nor global, but more or less long and more or less connected' (Latour 1993: 122).

To recognize the latter, however, is not to play down the significance of place and 'territoriality'. Most obviously, two decades of the application of radical political economy in development studies and urban and regional analysis serve to underscore that development, at whatever geographical scale, does not occur on the head of a pin, but unevenly in and through places. Thus, the globalized economy, too, has to be seen as a 'totality of interconnected processes and inquiries' (Wolf 1982) flowing through and combining in different ways in different local settings. One distinctive characteristic, as in the past, is the uneven distribution of tasks in the international division of labour to different locations offering specific attributes for capital accumulation. Another is the spatially differentiated assimilation and inflection of global imperatives, as the latter encounter places with distinct, historically layered, socio-economic structures and traditions. Globalization, therefore, does not imply a sameness between places, but a continuation of the significance of territorial diversity and difference.

Indeed, for some observers (e.g. Harvey 1989; Swyngedouw 1992), globalization, as the compression and transgression of time- and space-barriers, ascribes a greater salience to place, since firms, governments, and the public come to identify the specificity of localities (their workforces, entrepreneurs, administrations, and amenities) as an element for deriving competitive advantage. Place-marketing, in this context, is said to constitute a critical element, both for success in the interregional competition for investment (Lash and Urry 1992), and for global industry itself to derive competitive advantage and corporate distinctiveness (Ohmae 1989; Porter 1990). On a similar track, Watts has argued that 'globalization . . . revalidates and reconstitutes place, locality and difference' (Watts 1991: 10) because of the intense cross-cutting of multiple experiences, cultures, and images in individual places, yielding different hybrids of old and new influences in any given locality. Thus, the rise of global cultural flows, too, must be seen as a process of increased heterogenization at the local level, as 'cosmopolitan' cultures combine or clash (e.g. through migration) with previously rooted vernaculars.

In evaluating the salience of distinction and difference among places it has to be acknowledged that the mapping between the local and the global is not strictly isomorphic, but that the local has its own 'autonomy of being'. Johnston, asking why places continue to differ, comes to the conclusion that 'they do so because people make them so, not because of any necessary causal relationship but rather because of the spatially-varying nature of humanly-created milieux' (Johnston 1991: 137). Similarly, Pred, in defence of the locally spoken word and its political potential for 'turning the world upside down', writes that the 'historical unfolding of local civil society has a certain degree of autonomy' (Pred 1989: 218), and this, because of the 'locally singular combination of presences and absences, the locally peculiar sedimentation of practical and discursive knowledge, of commonsense, of behavioural dispositions and coping mechanisms' (ibid.). Thus, with or without globalization, local diversity, both within and between places, remains.

But, to emphasize the continued significance of local difference is not to suggest that nothing has changed in terms of the way in which places are constituted or relate to the rest of the world. Thus there is one influential account of contemporary change which anticipates a return to the region as the basic unit of economic, cultural, and political organization, as a result of the crisis of the national state-based system of capital accumulation, identify-formation, institutional representation, and political governance which dominated the post-war era. One story in this account refers to the rise of regions, such as Silicon Valley, Baden Württemberg, and the Third Italy, to predict the resurgence of specialized and self-contained regional economies and localized systems of governance and economic regulation that will replace the era of mass production and mass consumption which was based around the national economy and the nation state.

Another story draws upon the rise of diversified consumption norms, regional movements, and cultural demands, to anticipate a 'post-modern' future of multiple identities and plural social movements forged around local communities; an era in distinct contrast to the metanarratives of 'modernity' which were constructed around national evocations of class, gender, and race. In this 'back to the future' account of change, what is stressed is bounded local territorial dynamics which exercise a decisive influence on economy and society. Here, the global becomes a 'composition of local settings', separate and relatively self-governing regional social formations held together, as in the ages prior to modernization and the nation state, by place-based norms, identities, and rules of social behaviour (drawing upon common local economic activities, traditions, beliefs, and customs).

Our understanding of the local is different from this account of autarkic, rooted, and empowered localities constituted in and for themselves and therefore able to escape, in some measure, from global forces. Our disagreement is particularly with the concept of local 'separateness' (suggestive of the strength of local interconnections) and the hint of the local as an arena of homogeneity and authenticity in a world of inauthentic and unrooted global influences (Hebdige 1990). For us, globalization represents, above all, a greater tying-in and subjugation of localities (cities and regions) to the global forces described earlier. This is not to play down the significance of place and place-based phenomena, but to see localities as part of, rather than separate from, the global; the product of local, nationwide, and transnational influences. Nor is it to subscribe to an interpretation of the global as something of 'deep and decisive' influence and localities as something of less importance, as is suggested in a recent account which describes the global as 'structural', that is the arena of capital production and reproduction, and the local as 'contingent' (Taylor and Johnston 1991). In this account too, there is a tendency to see the local as separated from the global and subject to different sets of processes, instead of as a setting for both 'structural' and 'contingent' happenings.

We agree with Robins that 'it is important to see the local as a relational and relative concept, which once significant in relation to the national sphere, now . . . is being recast in the context of globalization . . . as a fluid and relational space, constituted only in and through its relation to the global' (Robins 1991: 35). However, we disagree with his claim that, as a result, the local should only be seen as a 'relational' rather than a 'real' space, that is not be mistaken for locality because of the territorial indeterminacy and lack of spatial fixity of the economic and cultural influences of today. It seems strange to us that a recognition of the local as 'in and of the global' should imply, as it does for Robins, an erosion of the local as bounded geographical space, as places with distinctive attributes, as recognizable cities and regions with their own 'physicality' and 'territoriality'. Why is it not possible to see globalization as a process affecting the basis upon which the identity of places is constructed, rather than as representing a transition to 'placelessness' or a challenge to 'place-boundedness'?

We therefore reject views of the local in the global as autarkic, 'contingent', amoeba-like, expressive of only physical fixity in the hyperspace of global capital flows (Swyngedouw 1992) or of identity and stability in the unsettling speed of modern existence (Harvey 1989). Instead, we wish to emphasize the continued salience of places as settings for social and economic existence, and for forging identities, struggles, and strategies of both a local and global nature. Almost a decade ago, one of us argued the case for conceptualizing regions as a medium and outcome of social interactions, as places in which lives are constituted and lived through, as arenas in which distinctive meanings are produced and contested (Thrift 1983). If we continue to accept a definition of territoriality as the basis for living in, assimilating, and making sense of the world, then there is no reason why globalization constitutes a threat to 'place'.

But, nor is there any reason, conversely, to cling on to a singular, pure, or homogeneous sense of place, of place-identification as a haven from complex and unwanted global intrusions (Alger 1990; Harvey 1989). For, as Massey argues, in reality, places do not have single pre-given identities, but are 'constructed out of the juxtaposition, the intersection, the articulation, of multiple social relations', and should be seen as 'shared spaces', 'riven with internal tensions and conflicts' (Massey 1991: 18–19). For Massey, around multiple meanings and identities cutting across any given locality, there still emerge valid, place-based movements drawing upon either place-bound traditions or wider symbols of liberation. The invitation, here, is to see the local as a real territorial arena of social interaction composed of difference and conflict, of related and unrelated connections, of social and economic heterogeneity, of parochial and universal aspirations, and of local and global determinations. But, and this is the point, each setting, with its distinctive sense of place and its 'bringing into play of locally sedimented, practice-based knowledge and experience, or the mobilization of collective memory, always produces a conjuncture of the local and the more global which is in some measure unique' (Pred 1989: 221) as well as 'a potential site of struggle and resistance' (ibid. 223). Place, then, has to be seen not just as a locus for 'being' but as a moment for 'becoming' (Massey 1993).

This account of the local in the local amounts to a recognition of place as both 'home' and 'the world'. It is a recognition of place-based, and sometimes place-bound, ways of living in the global. But it is not a fetishization of place or of local connections, or necessarily a celebration of the politics and policies of place against those operating at a wider scale. On the contrary, it is an acceptance of places as increasingly heterogeneous and disconnected arenas of social existence. For, as one observer reminds us, 'the story of global connections is also the story of disconnections . . . of people of different classes, and different sources of identity, within the same city and country' (Rosler 1989: 151).

Globalization, thus, represents a redefinition of places as juxtapositions of intersecting, overlapping, and unconnected global flows and historical fixities. It characterizes localities as territories living with different bits and pieces of the transnational division of labour as well as their own inherited industrial traditions; as territories composed of immigrant communities evoking imagined homelands, middle-class dwellers soothed by the 'authenticity' of distant exotic cultures, and working-class communities evoking a sense of place rooted in local traditions; and as territories selectively drawing upon rooted or imagined stories and myths to mobilize or subvert a common sense of regional identity (across real local social and economic divisions), in order to 'capture' the global (e.g. investment) or resist it (e.g. regionalism).

Increasingly, the pressure posed by globalization is to divide and fragment cities and regions, to turn them into arenas of disconnected economic and social processes and groupings. Nevertheless, these places continue to embrace singular and common identities in order to live in, or challenge, the global. The critical question which remains, then, is whether local strategies structured around common 'stories' and structures are actually successful in securing a place in the world, that is whether the politics and policies of place are appropriate or sufficient for securing acceptable levels of social and economic well-being within the global. Much of this question, of course, depends on where localities find themselves, in the first instance, within the global political economy, in terms of their established economic, political, and institutional strengths and weaknesses. In the next section, we examine the attributes of localities which do find themselves at the centre of global economic circuits, in order to initiate a discussion of the scope for retaining or developing local economic integrity in a global economy. For this reason, the discussion is largely restricted to questions of economic development.

HOLDING DOWN THE GLOBAL

The position of cities and regions in the global economy varies, of course, as does their window of opportunity on to it. In Europe alone, there are rich and poor agricultural areas offering the fruits of agro-business to the rest of the world and a meagre subsistence to its peasantry respectively. There exist old industrial areas which have continued to decline since the turn of the century as well as those that have undergone a more recent deindustrialization and are now seeking out a new role in the international economy. There are areas along the coastline and in the mountains, where old traditions and ways of life coexist with a tourism-based consumerism bringing riches to local businesses as well as multinational corporations. There are global cities which attract growth because of their strategic position in the international flow of money, goods, services, people, and images, but which also continue to shatter the dreams of millions of residents and migrants excluded from or marginalized by the economy. There exist also other, smaller, cities which rely on a mixture of urban services and functions to generate more modest but stable incomes for their citizens, and draw upon a pastiche of international architectural styles and local vernaculars to attract new investment and untapped consumer expenditure.[1]

These are examples we quote not only to illustrate differences in the way in which the local meets the global (and vice versa), but also to suggest spatial variability in the robustness and distribution of growth. The reasons for such variability remain far from clear, but one certainty is that globalization need not necessarily imply a sacrifice of the local. Indeed, the pertinent message to emerge from the volume of research on new industrial agglomerations such as Silicon Valley and industrial districts in Italy, or older centres of growth such as the City of London or Hollywood, is that global processes can be 'pinned down' in *some* places, to become the basis for self-sustaining growth at the local level.

Not coincidentally, in recent time local agglomeration has come to be treated in new ways as a result of the rise of a new institutional economics and a new institutional sociology. In particular, it is no coincidence that Marshall's notion of an 'industrial atmosphere' has been rediscovered, after a period in the 1960s and 1970s when industrial agglomerations were generally believed to exist purely as a means of easing the communication and control problems of large corporations. There are at least five clearly interrelated reasons for this rediscovery. First, research has shown that the industrial organization of the firm should be thought of in terms of loosely connected arrays, rather than as organic wholes. The firm is rarely as efficiently organized as the rhetoric of organization charts implies. Second, the tasks of coordination and control turn out to be nowhere near as straightforward as previously thought. They involve *ad hoc* decision-making and continual improvisation, often on a large scale. Third, firm-organization is often unstable. Certainly, it cannot be assumed that current organizational structures represent, in any sense, an efficient solution. Thus Powell and Dimaggio (1991: 33) can declare that:

we are sceptical of arguments that assume that surviving institutions represent efficient solutions because we recognise that rates of environmental change frequently outpace rates of environmental adaptation. Because sub-optimal organisational practices can persist for a long time, we rarely expect institutions simply to reflect current political and economic forces. The point is not to discern whether institutions are efficient, but to develop robust explanations of the ways in which institutions incorporate historical experiences into their rules and organising logics.

Fourth, the dividing line between firms and their environments is porous and constantly changes (Powell 1990). At least since the work of Emery and Trist (1972), the extent to which a firm can be seen as separate from its environment has been questioned. Just as in the social sciences the death of the subject has been announced, so in work on industrial organization there has been a corresponding death of the firm: or rather, it is now realized that there are many different firms caught up under one name, just as there are many subjectivities competing for attention in a human subject. Finally, many firms have clearly outgrown agglomerations. They are only ever partly present in such spaces and therefore very often they have both a global and a local identity, or more accurately their identity is a hybrid of the two, in which authority is no longer exercised by easily identified local élites but rather by shifting coalitions of global and local élites.

The literature on industrial agglomerations has generally moved towards a new approach based on a recognition of the importance of an institutional atmosphere in the creation and maintenance of agglomerations. At first, explanations were often couched in terms of the new institutional economics of Williamson (1975, 1985) and others, with their emphasis on transaction costs. However, it has become clear that the new institutional economics is radically undersocialized as an approach—a deficiency for which it paradoxically overcompensates by producing a radically oversocialized idea of society as a set of bureaucratic hierarchies in opposition to markets (Granovetter 1985). Thus more attention is now being paid to explanations couched in terms of a new institutional sociology, with its emphasis on 'embeddedness'.

'Embeddedness' refers to the 'fact that economic action and outcomes, like all social action and outcomes, are affected by actors' dyadic relations and by the structure of overall network of relations' (Grabher 1993: 4). Thus this approach stresses the importance of: seeing economic action as social action; understanding networks which function between markets and hierarchies on a semi-permanent basis; and tracking the processes of institution-building (Granovetter 1985). More recently, this approach has stressed the importance of particular common forms of understanding that are seldom explicitly articulated—classifications, routines, scripts, and other rationalizing and rationalized schemas or, in other words, institutional myths (Hodgson 1993).

The result is that attention in the literature on industrial agglomerations has increasingly turned from 'economic' reasons for the growth of new industrial agglomerations, such as product specialization and vertical disintegration of the division of labour, to 'social' and 'cultural' reasons such as intense levels of inter-firm collaboration; a strong sense of common industrial purpose; social consensus; extensive institutional support for local business; and structures encouraging innovation, skill formation, and the circulation of ideas (Hirst and Zeitlin 1991; Sabel 1989, 1992; Salais and Storper 1992; Storper 1993). The recognition of socio-cultural aspects has, in turn, given renewed impetus to the study of territorial embedded-ness as found in the literature on industrial districts and other localized production complexes.

However, the analysis of territorial embeddedness has only rarely been related in any systematic way to globalization. In a recent paper drawing upon experiences of the City of London and an industrial district in Tuscany (Santa Croce sull'Arno), we have attempted to identify some of the non-economic reasons why localized centres are still needed in the context of a globalization division of labour (Amin and Thrift 1992). Three reasons for place-centredness appear to be of particular importance in integrated global production filieres, serving to overcome problems of integration and coordination. First, centres provide face-to-face contact needed to generate and disseminate discourses, collective beliefs, stories about what world production filieres are like. They are also points at which knowledge structures can be tapped into. Second, centres are needed to enable social and cultural interaction, that is, to act as places of sociability, of gathering information, establishing coalitions, monitoring and maintain-ing trust, and developing rules of behaviour. Third, centres are needed to develop, test, and track innovations; to provide a critical mass of knowledgeable people and structures, and socio-institutional networks, in order to identify new gaps in the market, new uses for and definitions of technology, and rapid responses to changes in demand patterns.

These centres of geographical agglomeration are, thus, centres of representation, interaction, and innovation within global production filieres. They amount to 'intelligent regions' as illustrated in the chapter by Morgan and Cooke on Europe's most dynamic 'centred' economies. It is this unique ability to act as a pole of excellence and to offer to the wider collectivity, a well-consolidated network of contacts, knowledge structures, and institutions underwriting individual entrepreneurship, which makes a centre a magnet for economic activity. In other words, these centres are the forcing houses for the construction of worldwide contact networks through which discourses circulate and are modulated. But these networks do not all share the same qualities. They differ in the degree to which their members share the same backgrounds (e.g. class or gender or familial ties), institutional ties, and cultural and political outlooks. We might distinguish between at least three types of contact network, each one more tightly drawn than the former and each demanding a greater degree of trust. The first type consists of networks between members of different firms and between firms and major clients. These are often well cultivated by individuals on an enduring basis but they do not require their members to make major commitments of trust or to share values. A second type consists of networks of specialists who form interest groups, often bound together by particular professional–social institutions. Members of these networks will often have shared specialist knowledge, shared interests, and shared attitudes. Finally, 'epistemic communities' (Haas 1992) are tightly drawn interest groups which not only have shared specialist knowledge, shared interests, and shared attitudes, but also share a normative commitment to act in particular ways and to follow quite particular policy agendas.

What the study of these 'growth poles' illustrates only too clearly is that success—in terms of holding down the global (local embeddedness) and thereby generating self-reproducing growth—cannot be reduced to a set of narrow economic factors. This is not, of course, to claim that economic factors are unimportant—the experience of growth poles shows that a basic requirement for success seems to be the presence of strategic functions at the top end of any industrial division of labour or value-added chain. Instead, it is to claim that social and cultural factors also live at the heart of economic success and that those factors are best summed up by the phrase 'institutional thickness'.

Institutional thickness is a multifaceted concept. From the growing literature on the salience of institutions, it certainly seems possible to isolate the following factors that contribute towards the construction of institutional thickness. The first and most obvious of these factors is a strong institutional presence, that is a plethora of institutions of different kinds (including firms; financial institutions; local chambers of commerce; training agencies; trade associations; local authorities; development agencies; innovation centres; clerical bodies; unions; government agencies providing premises, land, and infrastructure; business service organizations; marketing boards), all or some of which can provide a basis for the growth of particular local practices and collective representations. However, although the number and diversity of institutions constitutes a necessary condition for the establishment of institutional thickness, it is hardly a sufficient one. Three further factors are also important (Powell and Dimaggio 1991).

Thus the second factor is high levels of interaction amongst the institutions in a local area. The institutions involved must be actively engaged with and conscious of each other, displaying high levels of contact, cooperation, and information interchange which may lead, in time, to a degree of mutual isomorphism. These contacts and interchanges are often embodied in shared rules, conventions, and knowledge which serve to constitute the 'social atmosphere' of a particular region. The third factor must be the development, as a result of these high levels of interaction, of sharply defined structures of domination and/or patterns of coalition resulting in the collective representation of what are normally sectional and individual interests and serving to socialize costs or to control rogue behaviour. Finally, there is a fourth factor. This is the development amongst participants in the set of institutions of a mutual awareness that they are involved in a common enterprise. This will almost certainly mean that there is a commonly held industrial agenda which the collection of institutions both depends upon and develops. This will usually be no more than a loosely defined script, although more formal agendas are possible. This agenda may be reinforced by other sources of identity, most especially various forms of socio-cultural identification (such as region, gender, and ethnicity).[2]

These four factors constitute a local institutional thickness defined as the combination of factors including inter-institutional interaction and synergy, collective representation by many bodies, a common industrial purpose, and shared cultural norms and values. It is a 'thickness' which both establishes legitimacy and nourishes relations of trust. It is a 'thickness' which continues to stimulate entrepreneurship and consolidate the local embeddedness of industry. It is, in other words, a simultaneous collectiv-ization and corporatization of economic life, fostered and facilitated by particular institutional and cultural traditions which appear to have been central to the generation of success within 'neo-Marshallian nodes in global networks' (Amin and Thrift 1992). In other words, what is of most significance here is not the presence of institutions *per se*, but rather the processes of institutionalization, that is the institutionalizing processes that both underpin and stimulate a diffused entrepreneurship—a recognized set of codes of conduct, supports, and practices which certain individuals can dip into with relative ease.

Brought together, these four determinants of institutional thickness will, in the most favourable cases, produce six outcomes. The first is institutional persistence, that is local institutions are reproduced. The second outcome is the construction and deepening of an archive of commonly held knowledge of both the formal and tacit kinds. The third outcome is institutional flexibility, which is the ability of organizations in a region to both learn and change. The fourth outcome is high innovative capacity, which is not just specific to individual organizations, but is the most common property of a region. The fifth outcome is the ability to extend trust and reciprocity. Finally, and least common of all, is the consolidation of a sense of inclusiveness, that is, a widely held common project which serves to mobilize the region with speed and efficiency.

In some nodes (e.g. the City of London) this favourable institutional thickness has persisted over a long period of time. In others (e.g. Silicon Valley) it might be argued that the maintenance of thickness has proved problematic, in particular because of the difficulty of controlling rogue behaviour in an individualistic cultural setting. It should also be remembered that institutional thickness in not always a boon. It can also be a trap, as shown by Glasmeier's chapter on the Swiss watchmaking industries' districts, and also, to some extent, by the emergent problems of Baden Württemberg as discussed in the chapter by Morgan and Cooke. It can produce resistance to change as well as an innovative outlook. Yet what does seem certain is that in those industrial contexts which are heavily reliant on the production of knowledge, innovation, and information for competitiveness, institutional thickness can have a decisive influence on economic development, notably in local situations in which such assets are not, as it were, 'locked into' individual firms. Then ideas, research capability, information, skills, specific assets, supply structures, and services made widely available through recognizable institutions such as centres of higher education, development agencies, business organizations, and other more informal associations, amount to the offer of specific and general infrastructures which can significantly ease the burden of entrepreneurship.

However, official attempts to produce what Sabel (1992) calls 'studied trust' in local agglomerations have proved problematic. It has proved relatively easy, through the establishment of various new institutions, to produce heightened levels of inter-institutional interaction but it has proved much more difficult to force a new collective representation or mutual awareness on these institutions, whether through coercive, mimetic, or normative means.[3] Yet what has become clear is that the economic success of localities in a global economy will increasingly depend upon the articulation between institutional thickness and 'economic' variables which make it worthwhile for industry to remain in a locality. Finding local economic integrity in a globalized world is, therefore, possible, but, if we take the institutionalizing processes identified above seriously, this suggests that the option is probably open to only a minority of localities with proven institutional capability.

CONCLUSION: INSTITUTIONAL THICKNESS AND LOCAL PROSPECTS

In this introduction we have tried to translate and develop the institutionalist paradigm of Granovetter (1985), Hodgson (1988, 1993), and others in the context of questions concerning local economic development, through the concept of institutional thickness. The definition of institutional thickness we have developed to capture the character of local agglomerations is broadly consistent with the concept of institution as deployed in recent critiques of the methodological individualism of neoclassical economies as made by Hodgson (1993) and others. In this literature institutions are very broadly conceived, to include not only formal organizations, but also more informal conventions, habits, and routines which are sustained over time (and through space). As such, institutions act to stabilize a range of collective economic practices in a particular territory. More generally, this means taking seriously the contention that the economic life of firms and markets is *territorially* embedded in social and cultural relations and dependent upon: processes of cognition (different forms of rationality); culture (different forms of shared understanding or collective consciousness); social structure (networks of interpersonal relationships); and politics (the way in which economic institutions are shaped by the state, class forces, etc.).

In other words, the paradigm is an attempt to look at institutions without assuming that because they have thick structures, they are necessarily functional in economic terms. The linkage between institutional thickness and economic development within Marshallian local economies is not intended to suggest that this, by definition, is a functional linkage and, as such, can be universally established in all locational contexts.

The need to problematize the relationship between institutional 'thickness' and economic development is particularly essential because of the uneven geography of local economic development prospects and possibilities. Three issues come immediately to mind. First, in geography, most of the literature on the relationship between institutions and economic development has focused on the anatomy of innovative regions or industrial districts, where the institutional milieu has played an obvious and positive role in underpinning economic success. But, does the particular prove the general? How, for instance, do we account for the success of areas such as the south-east of England, and, more particularly, the M4 corridor, where institutional 'thickness' of the sort described earlier has not been readily apparent? Or, conversely, what should we make of areas which are institutionally thick, but are not successful? The north-east of England, for example, as well illustrated in the chapter by Hudson, is overburdened by development corporations, proactive local authorities, business-related training initiatives, development agencies, government industrial development bodies, small-firm promotion ventures, and other institutions engaged in the business of promoting economic generation. The region also possesses a strong regional identity, common cultural aspirations, clearly identifiable business codes and rules of business conduct, and close contact between the actors who make up the region's social, political, and economic élite. Yet, the region can hardly be described as economically successful.

Institutionalists might argue that it is the regressive and conservative nature of the region's institutional 'thickness' which prevents economic restructuring. But then, the Marshallian areas too are hardly known to possess extroverted or progressive institutional arrangements. Looking at areas such as the north-east of England, the sceptic might be tempted to argue that institutional thickness, in its own right, does not promise economic adaptability or economic potential. Indeed, 'thickness' when expressive of a past economic trajectory may be positively an obstacle towards the institutionalization of new processes and structures appropriate for a different economic base. The chapter by Glasmeier and to some extent also that by Grabher, provide a good illustration of this problem.

A second under-explored issue within the institutionalist paradigm concerns the interface between globalization and institutional thickness at the local level. For example, it is often argued that local thickness influences the form that corporate plants and offices take in a particular place, but there is little work on the way in which specific corporate cultures influence local thickness or thinness. It may be possible that certain multinational companies take a more active role than others in establishing a presence in their region of location and extending this presence beyond simply the purchase of inputs, services, and skills. But, as argued in the chapter by Dicken, Forsgren, and Malmberg, such 'embedded' investments tend to gravitate towards the already institutionally rich metropolitan economies, while other local communities receive investors who at best might fund training colleges and cultural events, but fall well short of constituting a growth pole for the regional economy. This tension between rhetoric and reality is elaborated in different ways in the contribution by Robins and Cornford, who situate the construction of local cultures within global audiovisual spaces which have more to do with the imperatives of global media corporations than with economic processes with widespread local effects. The presence of local institutional thickness, thus, still need not amount to the fulfilment of the economic interests of the wider local collectivity.

A third open issue is whether local institution building of the sort we have seen in various 'neo-Marshallian nodes' is necessary for a locality to find a home in the global economy. In geography, the literature on institutional embeddedness and industrial districts nearly always assumes this to be the case, while the new institutional economics and sociology literature does not really concern itself with the spatiality of the structures which embed the economy. In some quarters of political science, however, there is a recognition that the increasing 'hollowing' out of the nation state, resulting from the crisis of the traditional functions of the Keynesian state and also the rise of international structures of governance, may be giving greater salience to both the local state and other local institutions of governance and economic regulation (Jessop 1992; Stoker 1990).

If local institution-building, however, is deemed necessary, then the prospects of many local economies in the advanced economies must surely look bleak. More often than not they are stripped of any coherent or cohesive institutional infrastructure and of any common cultural agenda, as a result of over a century of integration into the global capitalist project. In these localities, oppositional social movements against global instructions and other political projects of local empowerment, may not amount to much more than simply this, as suggested by Hudson in his account of north-east England.

Alternatively, in constructing an oppositional culture to vested and often moribund institutional interests, these movements and projects may unleash new socio-cultural forces, the consequences of which for the local economy cannot be predicted, as illustrated in the chapter on regionalism in southern Europe by Hadjimichalis. They may (or may not) unlock new mechanisms for attaining some form of local economic integrity within global economic networks. But at the very least, they represent a challenge to established processes of institutional conservation and conservatism which have long ceased to bring widespread economic benefits to the local community. On the other hand, it has yet to be convincingly illustrated that local economic (re)development requires *local* institutional thickness. Of course, its presence means that a locality is in a position to negotiate its position in the global economy. But, if it were possible to retain a context of strengthened national and supranational regulatory regime capable of safeguarding the interests of 'weaker' local economies, its salience might diminish. Ultimately, the question of local fortunes may be a matter of the sort of political choices which are made, rather than any pre-given necessity for structures of governance at the local level.

To conclude, there is clearly still some way to go before the ultimate power of the new institutional paradigm can be assessed as an explanation of geographically uneven development. In particular, it is apparent that in many cases we know very little about the institutional field of local areas; in some cases not even the most basic of institutional audits have been achieved. This is important because the institutional field of an area 'cannot be defined *a priori* but must be defined on the basis of empirical investigation. Fields only exist to the extent that they are institutionally defined' (Powell and Dimaggio 1991: 65). Still less can we claim to know much empirically about the strength or range of interactions between institutions in an area, the types of coalition that have resulted, or the construction of mutual awareness and common industrial agendas. This book therefore makes a first step in this direction through a set of industry and regional case studies each of which displays different levels of global engagement and local institutional capability.

NOTES

1. Of course, this logic also applies to the 'territoriality' of different industries. Some sectors are undoubtedly localized (e.g. certain kinds of high technology and craft industries). Equally, others have a more heterogeneous distribution.
2. e.g. Kanter (1977) has shown the importance of what she terms homosexual environments.

3. Coercive policy approaches are those in which institutions are effectively forced into networks through the imposition of procedures and rules. This can be the result of government action or the action of large firms on subsidiaries and subcontractors (e.g. in terms of quality). More likely, government- and firm-responses will attempt to mimic successful examples of institutional thickness. Finally, government- and firm-responses can be normative, as when certain standards of professional conduct are applied.

REFERENCES

Alger, C. A. (1990), 'Local responses to global intrusions', mimeo, Dept. of Political Science, Ohio State University.

Amin, A., and Thrift, N. (1992), 'Neo-Marshallian nodes in global networks', *International Journal of Urban and Regional Research*, 16/4: 571–87.

Appadurai, A. (1990), 'Disjuncture and difference in the global cultural economy', *Theory, Culture and Society*, 7: 295–310.

Bauman, Z. (1992), *Intimations of Postmodernity* (London: Routledge).

Castells, M. (1989), *The Informational City* (Oxford: Blackwell)

Chesnaux, J. (1992), *Brave Modern World: The Prospects for Survival* (London: Thames and Hudson).

Clifford, J. (1988), *The Predicament of Culture* (Cambridge, Mass.: Harvard University Press).

Dosi, G., Pavitt, K., and Soete, L. (1990), *The Economics of Technical Change and International Trade* (New York: New York University).

Emery, F., and Trist, E. (1972), *Towards a Social Ecology: Contextual Appreciations of the Future in the Present* (New York: Plenum).

Giddens, A. (1989), *The Nation State and Violence* (Cambridge: Polity).

—— (1990), *The Consequences of Modernity* (Cambridge: Polity).

—— (1991), *Modernity and Self-Identity* (Cambridge: Polity).

Gilpin, R. (1987), *The Political Economy of International Relations* (Princeton, NJ: Princeton University Press).

Grabher, G. (1993) (ed.), *The Embedded Firm: On the Socioeconomics of Industrial Networks* (London: Routledge).

Granovetter, M. (1985), 'Economic action and social structure: The problem of embeddedness', *American Journal of Sociology*, 91/3: 481–510.

Haas, P. M. (1992), 'Introduction: Epistemic communities and international policy coordination', *International Organisation*, 46: 1–35.

Hannerz, U. (1992), *Cultural Complexity* (New York: Columbia University Press).

Harvey, D. (1989), *The Condition of Postmodernity* (Oxford: Blackwell).

Hebdige, D. (1990), 'Fax to the future', *Marxism Today* (Jan.), 18–23.

Held, D. (1991), 'Democracy, the nation-state and the global system', *Economy and Society*, 20: 138–72.

Hirst, P., and Zeitlin, J. (1991), 'Flexible specialisation vs. post-Fordism: Theory, evidence and policy implications', *Economy and Society*, 20/1: 1–56.

Hodgson, G. M. (1988), *Economics and Institutions* (Cambridge: Polity).

—— (1993), *Economics and Evolution: Bringing Life Back Into Economics* (Cambridge: Polity).

Jessop, R. (1992), 'The Schumpeterian workforce state: On Japanism and post-Fordism', mimeo, Dept. of Sociology, University of Lancaster.

Johnston, R. J. (1991), 'A place for everything and everything in its place', *Transactions of the Institute of British Geographers*, NS 16: 131–47.

Kanter, R. M. (1977), *Men and Women of the Corporation* (New York: Basic Books).

King, A. D. (1991) (ed.), *Culture, Globalisation and the World System* (London: Macmillan).

Lash, S., and Urry, J. (1992), *Economies of Signs and Space: After Organized Capitalism* (London: Sage).

Latour, B. (1993), *We Have Never Been Modern* (Brighton: Harvester Wheatsheaf).

Massey, D. (1991), 'A global sense of place', Open University, D103 Block 6. *The Making of the Regions*, 12–51, Milton Keynes, Open University Press.

—— (1993), 'Power geometry and a progressive sense of place', in J. Bird, B. Curtis, T. Putnam, G. Robertson, and L. Tickner (eds.), *Mapping the Futures: Local Cultures, Global Change* (London: Routledge).

Metcalfe, J. S. (1988), 'Evolution and economic change', in A. Silberston (ed.), *Technology and Economic Progress* (London: Macmillan).

Ohmae, K. (1989), 'Planting for a global harvest', *Harvard Business Review*, 67/4: 136–45.

—— (1990), *The Borderless World* (New York: Harper).

Polanyi, K. (1944), *The Great Transformation* (New York: Rinehart).

Porter, M. (1990), *The Competitive Advantage of Nations* (London: Macmillan).

Powell, W. W. (1990), 'Neither market nor hierarchy: Network forms of organisation', *Research in Organizational Behaviour*, 12: 295–336.

—— and Dimaggio, P. J. (1991) (eds.), *The New Institutionalism in Organisational Analysis* (Chicago: Chicago University Press).

Pred, A. (1989), 'The locally spoken word and local struggles', *Environment and Planning D: Society and Space*, 7/2: 211–33.

Robertson, R. (1990), 'Mapping the global condition: Globalisation as the central concept', *Theory, Culture and Society*, 7: 15–30.

Robins, K. (1991), 'Tradition and translation: National culture in its global context', in J. Corner and S. Harvey (eds.), *Enterprise and Heritage* (London: Routledge).

Rosler, M. (1989), 'The global issue: A symposium', *Art in America* (July), 86 and 151.

Sabel, C. F. (1989), 'Flexible specialisation and the re-emergence of regional economies', in P. Hirst and J. Zeitlin (eds.), *Reversing Industrial Decline? Industrial Structure and Policies in Britain and her Competitors* (Oxford: Berg).

—— (1992), 'Studied Trust', Discussion Paper, Science Centre, Berlin.

Salais, R., and Storper, M. (1992), 'The four worlds of contemporary industry', *Cambridge Journal of Economics*, 16: 169–93.

Sassen, S. (1991), *The Global City* (Princeton, NJ: Princeton University Press).

Smith, A. D. (1990), 'Towards a global culture?', *Theory, Culture and Society*, 7: 171–91.

Stoker, G. (1990), 'Regulation theory, local government and the transition from Fordism', in D. King and J. Pierre (eds.), *Challenges to Local Government* (London: Sage).

Storper, M. (1991), 'Technology districts and international trade: The limits to globalisation in an age of flexible production', mimeo, School of Planning, University of California, Los Angeles.

—— (1993), 'Regional worlds of production: Learning and innovation in the technology districts of France, Italy and the USA', *Regional Studies*, 27: 433–55.

Strange, S. (1988), *States and Markets* (London: Frances Pinter).

—— (1991), 'An eclectic approach', in C. N. Murphy and R. Tooze (eds.), *The New International Political Economy* (Boulder, Colo.: Reinner).

Swyngedouw, E. (1992), 'The Mammon quest. "Glocalisation", interspatial competition and the monetary order: The construction of new scales', in M. Dunford and G. Kafkalas (eds.), *Cities and Regions in the New Europe* (London: Belhaven).

Taylor, P. J., and Johnston, R. J. (1991), 'Uneven development: General processes and local variations', mimeo, Dept. of Geography, University of Newcastle upon Tyne.

Thrift, N. (1983), 'On the determination of social action in space and time', *Environment and Planning D: Society and Space*, 1/1: 23–57.

Watts, M. J. (1991), 'Mapping meaning, denoting difference, imagining identity: Dialectical images and postmodern geographies', *Geografiska Annaler*, 73B/1: 7–17.

Williamson, O. E. (1975), *Markets and Hierarchies* (New York: Free Press).

—— (1985), *The Economic Institutions of Capitalism* (New York: Free Press).

Wolf, E. (1982), *Europe and the People without History* (Berkeley, Calif.: University of California Press).

2

The Local Embeddedness of Transnational Corporations

Peter Dicken, Mats Forsgren,
and Anders Malmberg

INTRODUCTION

Although the terminology may be recent, the question of the local embeddedness of transnational corporations (TNCs) has long been a central issue in the debate over the relationship between such organizations and the geographical areas in which they operate. As the geographical extensiveness of TNC activity progressively increased in the decades after World War II a broad consensus developed in much of the critical literature that TNCs were inexorably becoming 'placeless' organizations, owing no allegiance to country, region, or local community. They were seen as 'snatchers' rather than 'stickers', as operators of a metaphorical 'slash and burn' policy whereby areas were exploited for their resources (including labour) and then abandoned as the resource became exhausted or as better or cheaper alternatives were identified elsewhere. TNCs were seen as being intrinsically 'against' communities; the notion of 'capital versus the regions' (Holland 1976) gained wide currency. Implicitly, if not explicitly, this view of a harsh, geographically detached process was contrasted with a past, supposedly golden, age in which firms were both locally based and locally embedded in a socio-cultural, as well as an economic, sense; in which there was at least some identity of interest between firm and community; in which companies had a sense of obligation to their host communities. Such a view is, of course, very much a product of the use of rose-tinted spectacles. As Massey points out, 'even with geographical local ownership of production . . . profits could be withdrawn from "the region" and invested elsewhere. Regionally based capital is not necessarily regionally "loyal"; indeed, given that it is reasonably enterprising, it would be foolish to be so' (Massey 1984: 102).

Nevertheless, the stereotype view of the relationship between TNCs and geographical areas persisted. It was closely bound up with the Taylorist view of the organization of production and the Fordist regime of accumulation: the 'Old Competition' to use Best's (1990) term. However, recent developments in the organization and technology of production processes—the rise of flexible production—and in the emergence of the so-called 'entrepreneurial firm'—what Best terms the 'New Competition'—

have led to an emerging re-evaluation of the relationship between TNCs and local areas. The profound changes in the way products are made, how production is organized, and how firms relate both to their own internal members and to other firms—that is the alleged replacement of hierarchies by networks—are, it is claimed, changing the relationship between TNCs and localities. At the same time, TNCs themselves have become increasingly aware of the tension between globalizing and localizing forces. In other words, the 'local' is back on the agenda. Does the 'New Competition' hold out the potential of a positive reintegration of local areas within the strategies of TNCs in ways which enhance their developmental prospects, rather than diminish them? One of the aims of this chapter is to address this question.

In one sense, of course, all TNC activities are locally embedded simply because every activity must have a specific geographical location. That, in itself, means that something of the 'local' is 'contained' within each TNC unit. The extent to which this influences the behaviour of the individual unit or, indeed, of the TNC as a whole, is a subject of considerable current debate. The other side of the local embeddedness coin is the one more usually addressed in the debates alluded to above. To what extent are TNCs embedded in the local economic, social, political, and cultural networks which, together, define the essence of a particular locality? But this begs a further question: what do we mean by 'local' in this context? In the current geographical literature, 'local' generally refers to a very small geographical area, normally an individual city or urban region, although the precise definition and meaning of 'local' even in this sense is a contentious issue. To the TNC, however, 'local' may well mean 'national' or even 'regional' in the broader supranational sense.

One way of entering this debate on the local embeddedness of TNCs is through a lens which focuses upon the dynamic processes of evolving TNC strategies. The discussion is organized into two parts. The first part examines developments in the global organization of TNCs, specifically the dynamic diversity of TNC strategies and structures and the extent to which centralized hierarchical forms are becoming superseded by flatter, less-centralized network forms. The second part critically interprets the likely local impact of these complex developments. Specifically, it addresses the two dimensions of local embeddedness identified above. In particular, it questions whether we are simply witnessing the replacement of one stereotype by another or whether the new developments offer genuine developmental prospects for local areas which are, themselves, increasingly embedded within a global economic system. There is a third possibility, of course: that something new is happening but we do not really know, as yet, whether it is 'good' or 'bad' from a local perspective.

Three caveats need to be entered at this point. First, although we want to argue against the use of stereotypes we are often forced to use the term *the* transnational corporation. But this should not be taken to imply that we are slipping in a stereotype through the back door. Second, using the concept of 'strategy' as a point of entry in the analysis could give a misleading impression of objective rationality on the part of corporate decision-makers. Strategies, in fact, are invariably contested territory within TNCs and may be less self-conscious than they are made to appear. However, in the sense that strategy implies some form of broad, targeted, relatively long-term decision-making process then it does help to capture the essence of the kind of firm-behaviour discussed in this chapter. Third, the approach adopted is explicitly firm-centred: it examines the local from the perspective of the firm. It is also more heavily biased towards an economistic view, although implicit in our approach is the central fact that all economic behaviour is fundamentally socially embedded.

DEVELOPMENTS IN THE GLOBAL ORGANIZATIONS OF TNCS: FROM HIERARCHIES TO NETWORKS?

Until relatively recently, much of the business and management literature projected a linear trajectory of TNC development. This essentially teleological conceptualization was simple in the extreme: (1) a firm began as a producer for its domestic market; (2) it was then motivated to sell into export markets from its domestic base; (3) circumstances eventually encouraged the firm to establish local production facilities to serve its former export markets or to penetrate new markets; (4) technological changes especially in transport and communications made it increasingly possible for a firm to take advantage of geographical differentials in factor costs and availability on a global scale; (5) the firm is ultimately transformed into a fully integrated global corporation. As such geographical expansion occurs, the firm faces increasingly complex problems along two primary dimensions (Porter 1986). First, it has to decide where to locate the component parts of its production- or value-added chain (i.e. how to configure its operations geographically) and, second, it has to devise appropriate ways of coordinating its increasingly complex functional and geographical operations.

In both of these respects, again, a rather simplistic stereotype picture was painted, based primarily on the large US corporation. Not only was international direct investment in the post-war period dominated by US enterprises but also much of the early analysis of transnational corporations was based in the United States. In fact, there is—and always has been—very considerable diversity in TNC strategies and in the kinds of organizational coordination and geographical configuration employed to implement them (see e.g. Perlmutter 1969; Bartlett and Ghoshal 1989; Martinez and Jarillo 1989). Just as Perlmutter (1969), somewhat ahead of his time, was sufficiently sensitive to point to very substantial diversity among the 'multinational' firms of the 1960s, so too it can be argued today that the term 'global' corporation obscures a rich variety of actual forms. In addressing the question of the local embeddedness of transnational corporations the existence of such diversity must be kept constantly in mind.

TNCs can be categorized in a number of different ways in an attempt to capture something of this diversity. In Bartlett and Ghoshal's (1989) typology,[1] two major forces interact to produce different TNC structures: (1) the nature and complexity of the industry environment in which a firm operates and (2) the firm's own specific history (its administrative heritage, including the influence of national culture, which produces what Heenan and Perlmutter (1979) term its strategic predisposition). Bartlett and Ghoshal (1979: 49) identify three different organizational models, each with distinctive structural, administrative, and management characteristics: (1) the multinational organization model; (2) the international organization model; (3) the global organization model. The major characteristics of each of these three ideal types are summarized in Table 2.1.

The *multinational organizational model* emerged particularly during the interwar period. Firms were stimulated by a combination of economic, political, and social forces to decentralize their operations in response to national market differences. The multinational model is characterized by a decentralized federation of activities and simple financial control systems overlaid on informal personal coordination. The company's worldwide operations are organized as a portfolio of national businesses (Bartlett and Ghoshal 1989: 49). This was the model used extensively by expanding European companies. Each of the firms' national units had a very considerable degree of autonomy; each had a predominantly 'local' orientation.

The second type, termed by Bartlett and Ghoshal the *international organization model*, came to prominence in the period after World War II. It was typified by the large US corporations, many of which expanded overseas in the 1950s and 1960s to capitalize on their firm-specific assets of technological leadership or marketing power. This organizational model involved far more formal coordination and control by the corporate headquarters over the overseas subsidiaries. Whereas multinational organizations were, in effect, portfolios of quasi-independent businesses, international organizations clearly regarded their overseas operations as appendages to the controlling domestic corporation. Thus the international firm's subsidiaries were more dependent on the centre for the transfer of knowledge and information and the parent company made greater use of formal systems to control the subsidiaries (Bartlett and Ghoshal 1989: 50–1).

Bartlett and Ghoshal label their third organizational category the

TABLE 2.1. *Basic characteristics of different types of transnational business organization*

	Multinational	International	Global
Structural configuration	Decentralized federation. Many key assets, responsibilities, decisions decentralized.	Coordinated federation. Many assets, responsibilities, resources, decisions decentralized but controlled by HQ.	Centralized hub. Most strategic assets, resources, responsibilities, and decisions centralized.
Administrative control	Informal HQ–subsidiary relationship; simple financial control.	Formal management planning and control systems allow tighter HQ–subsidiary linkage.	Tight central-control decisions, resources, and information.
Management mentality	Overseas operations seen as portfolio of independent businesses.	Overseas operations seen as appendages to a central domestic corporation.	Overseas operations treated as 'delivery pipe-lines' to a unified global market.

Source: based on Bartlett and Ghoshal (1989: figs. 3.1, 3.2, 3.3).

'classic' *global organizational model*. This was both one of the earliest forms of international business (used, for example, by Ford and Rockefeller in the early years of the century) as well as a form used especially by Japanese firms in their much later internationalization drive of the 1970s and 1980s. This model is based on a centralization of assets and responsibilities. The role of the local units is to assemble and sell products and to implement plans and policies developed at headquarters. Thus, overseas subsidiaries have far less freedom to create new products or strategies or even to modify existing ones (Bartlett and Ghoshal 1989: 51).

Although each of these three ideal-type models developed during specific historical periods there is no suggestion that one was sequentially replaced by another. Each form has tended to persist, to a greater or lesser extent, producing a diverse population of transnational corporations in the world economy. There is some correlation between the type of organizational model used and the nationality of the parent company but, again, this cannot be regarded as a universal generalization. It is perhaps better to regard firms of particular national origins as having a predisposition towards one or other form of organization (see Heenan and Perlmutter 1979). We will explore the extent to which each of these organizational forms may produce different degrees and types of local embeddedness in the next section of this chapter. Before doing so, however, we need to examine what many argue is the current emergence of a totally new organizational form.

Quite clearly, the organization and managerial strategies of TNCs cannot be properly understood independently of the dynamics of the broader production system within which they operate. There is now a broad consensus, often from quite different ideological perspectives, that the global economy is undergoing dramatic, possibly revolutionary or epochal, changes. Developments in the technologies of products, processes, and organizational coordination; increasing time-space compression; changes in consumer markets; changes in international and national regulatory structures are all combining to increase the turbulence and uncertainty of the environment in which business firms, states, and localities exist. To some, indeed, the sea-change in the ways of organizing the production of goods and services and the restructuring of production systems, which intensified especially during the 1980s, indicate an all-embracing shift between two historical epochs: Fordism and post-Fordism (see e.g. Piore and Sabel 1984; Hirst and Zeitlin 1990; Storper and Scott 1989). Much of the impetus for this interpretation derives from the French regulationist school (for some pertinent discussion see Cooke *et al.* 1992; Moulaert and Swyngedouw 1989; Tickell and Peck 1992).

In this strand of literature it is assumed that a Fordist regime of accumulation (and its associated mode of social regulation) was, in some sense, hegemonic during the period from roughly the 1920s to the 1970s. Fordism had its physical base within an ensemble of mass production sectors such as cars, capital equipment, and consumer durables. These sectors were characterized by standardization of outputs, assembly-line methods, technical divisions of labour, and a rigid system of labour relations. Fully developed, 'this mode of social regulation comprised both the macro-economic steering mechanisms of Keynesian economic policy and the socially stabilizing influence of the welfare state' (Scott 1988: 173).

The deep economic recession triggered (although not fundamentally caused) by the oil crises from 1973 onwards, and which became manifest in the rise of Thatcherism in the United Kingdom and Reaganism in the United States, arguably signalled the beginning of the end for the Fordist hegemony. In place of mass production a number of more flexible forms of production were postulated which, it is argued, share a number of common features:

Unlike mass production activities . . . the new forms of production are generally characterised by an ability to change process and product configurations with great rapidity. . . . They are also typically situated in networks of extremely malleable external linkages and labour market relations. . . . One of the basic common traits of the flexible ensembles that have recently made their appearance in modern capitalism is their evident propensity to disintegrate into extended social divisions of labour, thus giving rise to many specialised subsectors. This process is a reflection of the tendency for internal economies to give way before a progressive externalisation . . . and leads at once to a revival of proclivities to locational convergence and reagglomeration (Scott 1988: 174–5).

The new regime of accumulation, then, whether it is referred to as post-Fordist, neo-Fordist, flexible specialization, or flexible accumulation is roughly characterized by: a shift towards customized goods; batch production; increased flexibility in various forms; a tendency to disintegrate vertically and, thus, to extend the social division of labour; and a tendency towards spatial (re-)agglomeration.

The debate over the extent to which an actual epochal shift has been, or is, occurring, together with its implications for the prospects for local economic development, is by now well rehearsed (see e.g. Gertler 1988, 1992; Hudson 1989; Amin 1989; Amin and Malmberg 1992; Lovering 1990). The emphasis on an abrupt historical transition is one of the basic sources of the critique of this new paradigm. Doubts have been expressed about whether it is at all possible to identify an unequivocal shift from Fordism (inflexible mass production of standardized goods within large, vertically integrated companies) to post- or neo-Fordism (flexible batch production of customized goods within small, vertically disintegrated, firms). Such doubts are twofold: do post-Fordist forms of production really dominate today and were Fordist forms of production really hegemonic in the past? Much of what has been written about 'Fordism's successors' seems to suffer from a huge overestimation of the importance of 'Fordist-like' organization of production during previous phases. Similarly, the predictions of some of the 'de-integrationists' that an increased externalization of business functions would lead to a dismantling of TNC structures run up against the empirical evidence of continued growth of TNCs during the 1980s and also the extent to which control of externalized functions may still be exercised by the coordinating firm (Cowling and Sugden 1987). We will not enter the thickets of these debates here. Rather we will simply note at this point that the 'post-Fordist scenario' provides an argument for increased local integration of production systems and we will return to it later in our discussion of the potential local embeddedness of TNCs.

The TNC-management literature itself has also become increasingly preoccupied by the challenges faced by business organizations—especially those operating across national boundaries—by the uncertainties of the new environment. Although terms like 'Fordism', 'post-Fordism', or 'modes of social regulation' are conspicuous by their absence in this literature (not surprisingly given its different ideological pedigree) there is a strong emphasis on the implications of the 'new flexibility' for organizational strategies and structures. Again, for our purposes it is useful to draw initially upon Bartlett and Ghoshal's analysis before looking more explicitly at the broader network paradigm.

Each of the three ideal-types of organization identified earlier possesses specific strengths but they also have some severe contradictions and tensions. Thus, the global company capitalizes on the achievement of scale economies in its activities and on centralized knowledge and expertise. But this implies that local market conditions tend to be ignored and the possibility of local learning is precluded. The more locally oriented multinational organization is able to respond to local needs but its very fragmentation imposes penalties for efficiency and for the internal flow of knowledge and learning. The international company 'is better able to leverage the knowledge and capabilities of the parent company. But its resource configuration and operating systems make it less efficient than the global company, and less responsive than the multinational company' (Bartlett and Ghoshal 1989: 58–9).

The dilemma facing firms—especially large firms—in today's turbulent competitive environment is that, to succeed on a global scale, they must possess three capabilities simultaneously. They need to be globally efficient, multinationally flexible, and capable of capturing the benefits of worldwide learning all at the same time. Rather confusingly, Bartlett and Ghoshal use the term 'transnational' to describe such an organization and write of the 'transnational solution'. Because of our preferred generic use of the term 'transnational' it is perhaps better to use the term 'complex global firm' for the newly emerging organizational form. (This does not necessarily imply an inevitable sequential development but rather that some firms are moving towards such a complex global structure.)

A key diagnostic feature of such organizations is their integrated network configuration and their capacity to develop flexible coordinating processes. Such capabilities apply both inside the firm (the network of intra-firm relationships which, it is argued, is displacing hierarchical governance relationships) and outside the firm (the complex network of inter-firm relationships). Hierarchical organizational forms are essentially vertically structured; each unit has a specific role defined by the controlling headquarters authority. The amount of horizontal interaction is invariably less than the amount of top-down, vertical interaction. In a network organization, in contrast, the interrelationships are predominantly horizontal rather than vertical; interdependencies are predominantly reciprocal (Thompson 1967).

There is now a rapidly growing literature on the network paradigm. Network relationships are seen as being central to the concept of the social embeddedness of economic action developed in the economic sociology literature (see e.g. Granovetter and Swedberg 1992; Zukin and DiMaggio 1990). In the business economics literature much of the initiative has been taken by Scandinavian researchers who interpret industrial systems as a whole, as well as the firms which constitute such systems, as networks (see e.g. Hagg and Johanson 1992; Hakanson 1989; Johanson 1989; Forsgren and Johanson 1992).

According to this view, the coordination of industrial systems is achieved neither through an organizational hierarchy nor through the price mechanism, as in a traditional market model. Rather, the central argument of the network approach is that coordination takes place through interaction among firms in the network. In such interaction, price or cost is but one of several influencing conditions. This view is consistent with the critique of transactions cost economics which questions the interpretation of firm-boundaries as the marginal cost interface between internalization and externalization (see Best 1990; Walker 1989; Cowling and Sugden 1987).

Interaction within economic networks takes the form of exchange processes which are similar to what Blau has characterized as social exchange relationships. Blau (quoted in Johanson and Mattsson 1987) describes the evolution of a relationship in the following way: 'Social exchange relations evolve in a slow process, starting with minor transactions in which little trust is required because little risk is involved and in which both partners can prove their trustworthiness, enabling them to expand their relation and engage in major transactions (Blau 1968, p. 454).' Relationships based upon a degree of trust rather than on administrative fiat, therefore, can take a long time to develop. This has profound implications for the ways in which business organizations behave, both internally and externally. One of the most significant developments of recent years, in this respect, has been the proliferation of a multiplicity of collaborative ventures between firms operating at an international scale. Strategic alliances, in particular, have become central to the competitive strategies of virtually all large (and many smaller) corporations. Such relationships are increasingly polygamous rather than, as in the past, monogamous. Their development reflects several aspects of the turbulent business environment including the high cost and risk of much new investment and problems in gaining access to certain markets (Business International 1987; Contractor and Lorange 1987; James 1985; Morris and Hergert 1987; Cooke 1988; Cooke *et al.* 1992; Wells and Cooke 1991). At the same time, the nature of customer–supplier relationships has been changing. Whilst international subcontracting has continued to develop both extensively and intensively (Holmes 1986) the introduction of just-in-time procurement systems within a broader system of production quality control reflects a more profound shift in the cooperative relationships between firms in networks. As Japanese companies indicate, a strong element of trust and mutual interdependence is central to the success of such relationships. As firms face increasing pressure to shorten their development cycles and to move from sequential to simultaneous relationships between design, production, and marketing functions (Stalk 1988; Stalk and Hout 1990; Schoenberger 1991) the need for reciprocally based interdependence in both intra- and inter-firm networks becomes a central concern.

Of course, the relatively flat nature of networks compared with hierarchical structures should not obscure the fact that there are significant differential power relationships within such networks. Not all parts of a business network are equal. Key roles are played by those firms which have the power to coordinate and perhaps even control assets and resources. Indeed, Cowling and Sugden (1987: 60) define a firm not in legal terms but as 'the means of coordinating production from one centre of strategic decision making'. Applied to the realm of business organizations operating across national borders, the generic transnational corporation can no longer be defined simply in terms of equity relationships and foreign direct investment data become even less reliable as a real indicator of the scale and distribution of transnational activities. The modalities of transnational involvement are extremely diverse. They may be seen as an indication of

the distinction between Fordist multinational organisational forms and those of post-Fordist globalised organisational forms. The globalised firm has indistinct and shifting boundaries, we may expect it to be networked or distributed in organisational structure rather than being hierarchical, and it may penetrate and exploit space by proxy or in cooperation with other firms rather than in 'isolation' (Wells and Cooke 1991: 17).

Applying a network approach to the analysis of TNC configuration and coordination means treating business relations between firms and corporate relations within TNCs in much the same way. Thus, a subsidiary is part of a network that includes other actors outside as well as inside the corporation. The strategic position of the individual corporate unit develops over time in interaction with other actors in the network as much as by any attempt of corporate management to design the organization in a particular way. A strategic network position can be used by any unit to influence the behaviour of other corporate units and sometimes, indeed, the overall strategy of the TNC itself. This is not to say that the hierarchy and the actions taken by top management are of little or no importance. What we do argue, however, is that the attempt from the top of the hierarchy to implement an overall corporate strategy is but one force amongst others which influence the configuration of TNC activities.

The efforts of top management to coordinate are counteracted by the involvement of the subsidiary in business activities based upon successive and reciprocal adaptations to other units within the network. If we accept that the control over critical resources is an important power base, and that such control is related to network position, then we can conclude that certain subsidiaries will exert considerable influence on, sometimes in opposition to, the interests of top management. For example, in the case of a reduction in the production capacity of a foreign subsidiary the tension between its position in the network and the intentions of top management often emerges clearly. The subsidiary and local institutional bodies often appear as allies against corporate management. It can generally be assumed that the longer the subsidiary has been operating in a country or region, the more its behaviour will be influenced by locally embedded network relationships.

To claim that corporate relationships are 'networked' in much the same way as inter-firm relationships does not imply that the relationships are uniform in shape or character. It would be a mistake, therefore, to assume that all firms are proceeding down the same 'network paradigm' path (Cooke and Morgan 1991). What we have in reality is a variety of developmental trajectories and a spectrum of forms of corporate govern-ance with differing degrees of networked interrelationships of power and influence (Dicken and Thrift 1992). Nevertheless, an increasingly widely heard clarion call in executive boardrooms and in company annual reports is that of 'global localization', especially within the global Triad of North America, Western Europe, and East and South East Asia (Ohmae 1985). Such a strategic thrust clearly has potentially profound implications for the local embeddedness of TNCs and their operations.

INTERPRETING THE 'LOCAL' WITHIN THE GLOBAL DYNAMIC OF TNC DEVELOPMENT

In the apparently rapidly globalizing world of the TNC does the 'local' still have a place? Much depends, as suggested earlier, on what is meant by 'local'. In the international business world, 'local' is invariably equated with 'national'. At that scale, two very different views currently exist. Writers such as Ohmae (1990) and Reich (1991) argue that a corporation's nationality is increasingly irrelevant both to the corporation itself and to the state. Porter (1990), on the other hand, adopts a very different, and more geographically sophisticated, position. In his view,

The conditions that underlie competitive advantage are indeed often localised *within* a nation, though at different locations for different industries . . . But nations are still important. Many of the determinants of advantage are more similar within a nation than across nations . . . it is the *combination* of national and intensely local conditions that fosters competitive advantage (Porter 1990: 157–8).

Porter has obviously discovered geography (though not necessarily the work of geographers). He makes much, in his analysis of the 'competitive advantage of nations', of the importance of the local milieu, of the positive, dynamic, innovative benefits of spatial agglomeration. Without referring specifically to it he echoes much of the geographical work on 'new' industrial spaces.

At least in origin, TNCs are 'locally grown'; they develop their roots in the soil in which they were planted. The deeper the roots, the stronger will be the degree of local embeddedness such that they should be expected to bear at least some traces of the economic, social, and cultural characteristics of their home territory. In other words, they continue to contain elements of the local within their modes of operation. It is often argued that as a firm becomes increasingly transnational and, perhaps, global, it becomes placeless. But the evidence suggests that this is wrong. It is significant, for example, that the three different types of transnational organization discussed in the previous section are, at least partly, associated with different national origins. Chandler (1986) has shown how different organizational forms and distinctive managerial attitudes evolved in different national circumstances. Dunning (1979) relates a series of firm-specific attributes to the home country environment in which they are based.

All this, together with much anecdotal evidence, lends support to the view that the local socio-cultural milieu is a major influence on how firms evolve and behave even when their operations are geographically very extensive. This is not to argue a case for cultural determinism or even to argue that all TNCs of a given nationality are identical. Clearly, they are not. But they do tend to share some common features. Such commonalities can be regarded as a reflection of 'embeddedness . . . [as] . . . the contingent nature of economic action with respect to cognition, culture, social structure and political institutions' (Zukin and DiMaggio 1990: 15).

Although the question of the influence of the local milieu on TNC behaviour is important and merits further research, it is the extent to which TNCs *affect* the localities within which they operate which is the real focus of concern in this chapter. There has been much discussion of TNC impact on home and host economies (for a review, see Dicken 1992: ch. 12; Dicken *et al.* 1994). From the perspective of 'embeddedness', however, one of the major issues would seem to be the extent to which TNCs do, or do not, participate in local economic and social networks. It is this question which is addressed in the remainder of this chapter.

We noted earlier that TNCs have to make strategic decisions along two specific, but related, dimensions. First, they have to decide where to locate specific operations and, second, how to coordinate them. Invariably, of course, TNCs do not grow as the precise outcome of carefully thought-out strategies based upon a blank map of the world. TNCs grow, and develop, incrementally, often haphazardly and frequently by absorbing existing companies' operations through acquisition and merger. Configurational and coordinative decisions have to be made, therefore, on the basis of existing structures whose parts were incorporated at different periods of time and, often, in very different circumstances. Hence, many of the TNCs' decisions along the two dimensions are part of a continuing process of restructuring and not merely simple geographical expansion. The questions are most commonly ones of how existing facilities should be used to meet changing strategic needs: how production chain functions should be allocated and reallocated between existing sites: how new coordinating mechanisms should be applied in the face of an existing administrative structure.

For a particular geographical area, then, there would seem to be two major determinants of the extent of local embeddedness. First, the type of TNC operation involved, that is its position within internal production chains and external business networks, as well as the kinds of functions it performs, is clearly important. There is a large and well-known literature concerned with this issue which focuses particularly on the varying geographical orientations of the major corporate functions: headquarters, research and development, and manufacturing, in particular. Much of this work was inspired by Hymer's (1972) assertion that the internal division of labour within a TNC was projected directly on to geographical space to produce a parallel international spatial division of labour (see e.g. Dicken 1992; Massey 1979, 1984; Lipietz 1980). Although this conceptualization has, rightly, been criticized for its oversimplification and reductionism (Sayer 1985; Walker 1989) it contains some elements of truth. Both internationally and intranationally a corporate spatial division of labour has developed, although it is far more complex and dynamic than Hymer's model would have us believe. But it is certainly true that different elements within the production–value-added chain do have different local implications, different potentials for local embeddedness, although it is not possible to read off from this characteristic alone.

The second, and most important, determining factor is the way in which the TNC tries to coordinate its operations and especially the degree and kind of influence units have on corporate strategic behaviour. Important contributing factors here would seem to be the history and national origin of the TNC itself, the size of the units, the 'corporate importance' of the assets and resources that the units control, and the degree and kind of autonomy and influence that the TNC units have. The latter is, in turn, partly determined by the history and national origin of the TNC itself.

Nevertheless, there is considerable variety, even between TNCs with similar corporate histories and identical national origins. For example, the two large Swedish companies SKF and Ericsson both have an extensive record and a long history of internationalization, with production operations in many different countries. In both cases, the degree of local embeddedness is high in terms of business relations with local suppliers, customers, and institutions. But in the case of SKF each foreign subsidiary is specialized in certain products and there are extensive flows of goods between various corporate units. In this way, each individual SKF unit has important direct and indirect links to corporate units located in other countries. This is not so in the case of Ericsson, in which each subsidiary produces only for the local market. In SKF, marketing is very much standardized whilst Ericsson's subsidiaries adapt their sales and service to the specific requirements of the local customers. On the other hand, product development seems to be more decentralized in SKF than in Ericsson.

The importance of the local arena is even more pronounced in a third Swedish transnational company, AGA. The characteristics of the product (gas applications for industrial use) and the production process delimit the spatial extension of the business environment of each plant to the local area. Contrary to the Ericsson case, in which the national border is the most important demarcation line for production and marketing activities, AGA's plants are always located in immediate proximity to a customer or a cluster of customers. An AGA plant may well serve customers in several countries as long as they are located close to each other. Thus, for example, the plant in the Ruhr area serves customers in the Benelux countries as well as in Germany. Marketing and product development are carried out by each subsidiary and adapted to the needs of local customers with very little integration between subsidiaries in different countries.

Perhaps the degree of local embeddedness should be assessed by 'function' rather than at an overall 'corporate level'. For example, in Sweden's largest pharmaceutical company, Astra, most of the production is located in Sweden and the products are exported to more than one hundred countries. But this strongly concentrated production system is combined with a high degree of local adjustment in marketing. Astra has sales subsidiaries in several countries, performing their selling activities in close cooperation with local customers and health authorities. Thus, the degree of local embeddedness may differ, not only between companies but also within them, depending upon the function or activity in question. Within one individual company, a highly globalized activity with important relationships outside the local area may be combined with another activity having predominantly short-distance business relationships. Production may be either localized or globalized, but this is equally true for purchasing, marketing, R&D, and so on.

A more general approach to these questions is through the use of the major ideal-types of TNC organization discussed in the previous section of this chapter. Recall that each of the types identified by Bartlett and Ghoshal (and by others working along similar lines—see Martinez and Jarillo 1989) tends to employ different methods of administrative control and to regard their overseas subsidiaries in quite different ways. Allowing for the simplifications inherent in such rather coarse typologies they do help to make the point that even during the years when all TNCs were alleged to conform to the stereotype image of total centralization of control and negligible local sensitivity there was, in fact, substantial variation. In particular, the large number of TNCs, especially from Europe but not exclusively so, which operated (and continue to operate in some cases) a 'multinational' organizational form gave their local subsidiaries a substantial degree of local autonomy. Companies operating a simple global strategy based upon centralized production and those operating a more formally structured 'international' type of organization presumably give least autonomy to their local subsidiaries. But even in the latter category there may be considerable variation between companies, even in the same industry. In each case, however, we would expect that the longer a subsidiary has been operating in a particular country or region the more its behaviour would be embedded locally.

The shift towards more complex global organizations is resulting in a major re-evaluation by many firms of the roles to be played by their subsidiary units and the ways in which they are to be coordinated. In the face of intensifying global competition, rapid technological change, and changing political pressures at national and supranational levels, TNCs are restructuring their activities in ways which have profound implications for local areas. As with the debate over the pervasiveness of flexible specialization in general it is important to avoid exaggerating the extent to which the older TNC forms are being displaced by the newer, more flexible, forms of organization. Nevertheless, the boundaries between firms are in flux and are becoming more blurred as the interface between the internalization and externalization of functions shifts. It is now clearer than ever that TNC units are embedded in two different networks: they are embedded within their internal corporate networks and they are embedded within an external business and institutional network. The critical question from the point of view of this chapter is the extent to which this external network is geographically local or non-local. The two major influences here would seem to be, first, parent company policy towards the externalization of key functions and, second, the extent to which a particular local environment meets the perceived requirements of the TNC.

The 'fit' would seem to be closest in the case of the TNC's headquarters operations which, it is generally agreed, tend to be very strongly embedded within their local milieu. Historically, the fit has been weakest in the case of the 'branch plant' operation. But, as we have argued, times are changing. We should not persist unthinkingly with the branch plant stereotype but be prepared to acknowledge that not all branch plants, or other kinds of TNC unit, are the same. The local impact and embeddedness of such activities has to be set within the precise roles performed within the corporate network. Again, Bartlett and Ghoshal (1986, 1989) make a useful contribution in this respect. Their argument is that TNCs, particularly the complex global network forms, are increasingly differentiating the roles and responsibilities of their national subsidiaries according to two related dimensions: the strategic importance of the local environment to the firm on the one hand and the level of local responsibility and capabilities of the subsidiary on the other. Using this framework they identify four kinds of subsidiary role: the strategic leader, the contributor, the implementer, and the black hole. Each of these differs in their contribution to overall corporate strategy and may impact upon local economies in different ways. Most local subsidiaries remain essentially as 'implementers' of centrally decided policy but, increasingly, some local subsidiaries are being given more strategic roles. Not only is their position in the corporate network being enhanced but also it might be expected that their local embeddedness may be increased.

From an economic perspective, probably the most important single indicator of local embeddedness relates to supplier relationships. Two related dimensions are involved here (Fig. 2.1). First, there is the extent to which the local subsidiary is free to choose its suppliers (i.e. the degree of local autonomy in purchasing) and, second, the extent to which such choice is exercised locally. There is widespread agreement that major changes have been taking place in the relationships between firms and their suppliers in recent years as part of a redefinition of how production is organized within and between firms. The conventional wisdom associated with the 'flexspec' school is that, as an increasing proportion of a firm's production chain functions are externalized and as organizational forms take on a flatter network form, the opportunities for 'local' suppliers to do business with TNCs will increase. A further argument is that, as TNCs increasingly utilize just-in-time procurement systems they will be encouraged to source locally to minimize time delays. We should be aware, however, of two important points. First, not all production systems are likely to move in the same direction and, second, that there is no necessary connection between the two dimensions (axes). It is quite conceivable that increased organizational autonomy for a TNC subsidiary may lead to globalization of purchasing.

In fact, time has become a major focus of concern within business firms; it is, in Stalk's (1988) view, 'the next source of competitive advantage'. But this involves a great deal more than mere geographical proximity. Indeed, neither Stalk (1988) nor Stalk and Hout (1990) are concerned with distance at all. To them, the time problem is one that pervades the entire production process, from initial conceptualization to final marketing, and involves the compression of product development cycles. Schoenberger (1991), however, does address the specific issues of time and distance in relation to control within multinational firms. The spatial concentration of customers and suppliers should promote efficiencies and specialization and enhance the innovation process. This notion is at the heart of Storper's

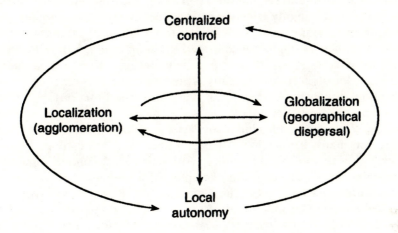

Fɪɢ. 2.1. *Dimensions of the global–local relationship*

(1992) 'technology districts' and, indeed, many of the 'new industrial districts' identified in the geographical literature in recent years. This would seem to suggest an increased incentive for TNCs to become more locally embedded than in the past when the geographical extent of supply networks became global and many firms bought primarily on the basis of price.

The new customer–supplier relationships involve longer-term, closer relationships based upon a high level of mutual trust; they are deeply socially, as well as economically, embedded. As such, they offer much greater potential for firms of all sizes to engage in integrated network activities. But a corollary of such closer, deeper relationships is that the supplier population becomes increasingly differentiated. Many major firms now operate an upper tier of 'preferred suppliers' who are closely integrated at all stages of the production process, from design to final production. For any one firm, such preferred suppliers will be relatively few in number. Not every local economy, therefore, can hope to participate in these new integrated networks. The smaller the geographical area in question the less likely it is to possess supplier firms of the necessary quality. Local embeddedness, therefore, is a double-sided coin: it reflects both the choices of the TNC and the existence of appropriate firms with which they can interact.

More than this, the trends which are occurring in the geographical distribution of TNCs' production–value-added chain elements suggest that the benefits will be distributed very unevenly. Although most firms have been restructuring their operations and several have begun to disperse some of their headquarters functions (for example, IBM, Hewlett Packard, Nestlé) dispersal is highly selective geographically. The same applies to the geographical rearrangement of R&D facilities. Both of these 'high level' functions remain strongly concentrated geographically, primarily in the firms' home country or region precisely because they are so deeply embedded locally. It is certainly true that the concept of 'global localization' involves greater local integration of functions but this is predominantly at the scale of the three major global regions: North America, Western Europe, Japan and East and South East Asia. As noted earlier, for many firms, 'local' has a rather broad meaning. Such an interpretation is reinforced by the way in which local content regulations are operated. In the EC case, for example, local sourcing is defined as that which occurs anywhere within the Community and not at the level of the individual country, let alone that of the city or local region.

The undoubted trend towards 'localization', therefore, needs to be kept in perspective. As McGrath and Hoole (1992) demonstrate, the 'globally localized' corporation is still, essentially, global. Being more sensitive to 'local' differences does not necessarily make the TNC more locally embedded in a real sense. It may well do so in some cases but, even in such cases, it is clearly a case of globalized local embeddedness.

Schoenberger (1991) aptly summarizes the likely regional–local implications of the new forms of production organization based upon integrated networks involving complex inter-firm collaborations:

First, the most likely scenario is that a smaller number of places will become the hosts to more integrated multinational corporate investments. For these favored places, the prognosis is relatively good as the high level of integration will yield a more diverse and qualified occupational structure. Moreover, the stability of the investment, implicating as it does multiple linked firms, is likely to be significantly higher. Yet, while many local firms will no doubt be drawn into the production complex, it is perhaps less likely that they will become core members of the collaborative partnership, remaining rather in a subordinate position to it. Secondly, as these investments become more concentrated in particular regions, the excluded regions are likely to become that much more excluded. Rather than a gradual embracing of more and more territory into the productive orbit of multinational networks, the degree of geographical differentiation in terms of investment, growth, and sectoral mix will tend to increase (Schoenberger 1991: 21–2).

CONCLUSION

From the viewpoint of inter-firm relationships, therefore, the prospects for greater local embeddedness of TNCs created by the new organizational forms appear to be limited to a minority of favoured places. Just as the old stereotype of 'placeless' TNCs needed to be revised to recognize that there has always been great variety in TNC–local relationships so, in turn, the new stereotype of increasing local embeddedness on a wider scale also needs to be viewed in a critical light.

There are, of course, other aspects of local embeddedness which have not been addressed in this chapter. This is partly because of shortage of space but also because of a paucity of empirical evidence. For example, there is anecdotal evidence that some TNCs do make great efforts to involve themselves in their local communities through such ventures as the sponsorship of local sporting and cultural activities, through direct involvement in training programmes and so on. But it is impossible to get a comprehensive picture of this kind of social embeddedness or even to assess whether it is deep or merely cosmetic. Do some types of TNC make a greater attempt to build local relationships than others? It is notable, for example, that Japanese firms in the United Kingdom have expressed strong views on what they see as their obligations to the local communities in which they operate. Is this a cultural trait of the Japanese? Is it a calculated move to deflect criticism? The German automobile components manufacturer Robert Bosch is currently funding a community training institution in South Wales where it has recently made a major new investment. Is this narrow self-interest or a reflection of community spirit—or both? To what extent do TNCs endeavour to employ local people in senior management positions? Many firms clearly have a policy to do that; others tend to move expatriate workers around their international operations and then try to present this as 'local' participation. For example, a recent advertisement for staff by BAT Industries reads as follows:

Our 46 companies around the world are not managed by remote control. They are run by people who live in the local community, understand the local culture and speak the language.

Every few years these people move to new parts of the world and begin afresh, developing their careers and making themselves at home in environments as diverse as Africa, Asia, the Far East, the Americas and Europe (*Financial Times* (2 July 1992)).

A final question which needs to be addressed is this. If it is regarded as desirable to try to increase the local embeddedness of TNCs then what can local institutions do to help? In the United Kingdom, the Business in the Community Programme endeavours to stimulate relationships between business firms and their local communities. With few exceptions, virtually all local government authorities make strenuous efforts to attract TNC investments into their territories. Yet the evidence from a comprehensive survey of local authorities in northern England shows that virtually none of them goes any further and engages in 'after-care' or 'after-sales service' (Dicken and Tickell 1992). In the final analysis it may be that, because of the immense asymmetry of power between TNCs and local institutions, there is little that local authorities can do other than to try to provide an attractive 'business environment' in which TNCs can operate, including, perhaps, the stimulation of appropriate local businesses which might be eventually embedded within a TNC's network. However, to avoid the dangers of mismatch this, itself, requires a close working relationship with both sets of firms, TNC and local. Ultimately, virtually all effective bargaining power lies, not at the local level, but at the level of the national state. To that extent, the nature and degree of local embeddedness may well be influenced as much by national policies as by local actions.

NOTE

1. Bartlett and Ghoshal's work reflects some of the terminological confusion which exists in the literature on international business organizations. They use the term 'transnational' organization to refer to a specific organizational form which, they believe, represents a particular solution to the problems of modern global competition. The United Nations and many others use the term 'transnational' in a generic sense to refer to all business which operates across national boundaries in ways which go beyond simple trade.

REFERENCES

Amin, A. (1989), 'Flexible specialization and small firms in Italy: Myths and realities', *Antipode*, 21: 13–24.
—— and Malmberg, A. (1992), 'Competing structural and institutional influences on the geography of production in Europe', *Environment and Planning A*, 24: 401–16.
Bartlett, C. A., and Ghoshal, S. (1986), 'Tap your subsidiaries for global reach', *Harvard Business Review* (Nov.–Dec.), 87–94.
—— (1989), *Managing across Borders: The Transnational Solution* (Boston: Harvard Business School Press).
Best, M. H. (1990), *The New Competition: Institutions of Industrial Restructuring* (Cambridge: Polity).
Business International (1987), *Competitive Alliances: How to Succeed at Cross-Regional Collaboration* (New York: Business International Corporation).

Chandler, A., Jun. (1986), 'The evolution of modern global competition', in Porter (1986).

Contractor, F., and Lorange, P. (1987) (eds.), *Cooperative Strategies in International Business* (Lexington, Ky.: Lexington Books).

Cooke, P. N. (1988), 'Flexible integration, scope economies and strategic alliances: Social and spatial mediations', *Environment and Planning D: Society and Space*, 6: 281–300.

—— and Morgan, K. (1991), 'The network paradigm: New departures in corporate & regional development' (Regional Industrial Research Report, 8; Cardiff).

—— Moulaert, F., Swyngedouw, E., Weinstein, O., and Wells, P. (1992), *Towards Global Localization* (London: UCL Press).

Cowling, K., and Sugden, R. (1987), 'Market exchange and the concept of a transnational corporation', *British Review of Economic Issues*, 9: 57–68.

Dicken, P. (1992), *Global Shift: The Internationalisation of Economic Activity* (2nd edn., London: Paul Chapman Publishing: New York: Guilford Publications).

—— and Thrift, N. J. (1992), 'The organisation of production and the production of organisation: Why business enterprises matter in the study of geographical industrialization', *Transactions of the Institute of British Geographers*, NS, 17: 279–91.

—— and Tickell, A. T. (1992), 'Competitors or collaborators? The structure of inward investment promotion in northern England', *Regional Studies*, 26: 99–106.

—— Quevit, M., Savary, J., Desterbecq, H., Nauwelaers, C. (1994), 'Strategies of transnational corporations and European regional restructuring: some conceptual bases', in P. Dicken and M. Quevit (eds.), *Transnational Corporations and European Regional Restructuring* (Utrecht: Netherlands Geographical Studies).

Dunning, J. H. (1979), 'Explaining changing patterns of international production: In defence of the eclectic theory', *Oxford Bulletin of Economics and Statistics*, 41: 269–96.

Forsgren, M., and Johanson, J. (1992), 'Managing in international multi-centre firms', in eid. (eds.), *Managing Networks in International Business* (Philadelphia: Gordon and Breach).

Gertler, M. S. (1988), 'The limits to flexibility: Comments on the post-Fordist vision of production', *Transactions of the Institute of British Geographers*, NS, 13: 419–32.

—— (1992), 'Flexibility revisited: Districts, nation-states, and the forces of production', *Transactions of the Institute of British Geographers*, NS, 17: 259–78.

Granovetter, M., and Swedberg, R. (1992), *The Sociology of Economic Life* (Boulder, Colo.: Westview Press).

Hagg, L., and Johanson, J. (1992) (eds.), *Foretag i nätverk [Firms in Networks]* (Stockholm: SNS Förlag).

Hakanson, H. (1989), *Corporate Technological Behaviour, Cooperation and Networks* (London: Routledge).

Heenan, D. A., and Perlmutter, H. (1979), *Multinational Organisational Development: A Social Architecture Perspective* (Reading, Mass.: Addison-Wesley).

Hirst, P., and Zeitlin, J. (1990), 'Flexible production versus post-Fordism, evidence and policy implications' (Birkbeck Public Policy Centre Working Paper; Birkbeck College, University of London).

44 *Local Embeddedness of Transnational Corporations*

Holland, S. (1976), *Capital versus the Regions* (London: Macmillan).
Holmes, J. (1986), 'The organisation and locational structure of production subcontracting', in A. J. Scott and M. Storper (eds.), *Production, Work and Territory: The Geographical Anatomy of Industrial Capitalism* (Boston: Allen and Unwin).
Hudson, R. (1989), 'Labour-market changes and new forms of work in old industrial regions: Maybe flexibility for some but no flexible accumulation', *Environment and Planning D: Society and Space*, 7: 5–30.
Hymer, S. H. (1972), 'The multinational corporation and the law of uneven development', in J. N. Bhagwati (ed.), *Economics and World Order* (New York: Free Press).
James, B. G. (1985), 'Alliance: The new strategic focus', *Long Range Planning*, 18: 76–81.
Johanson, J. (1989), 'Business Relationships and Industrial Networks', in *Perspectives on the Economics of Organisation* (Crafoord Lectures, 1; Institute of Economic Research, Lund University Press).
—— and Mattsson, L.-G. (1987), 'Inter-organizational relations in industrial systems: A network approach compared with the transaction cost approach', *International Studies of Management and Organization*, 17: 34–48.
Lipietz, A. (1980), 'The structuration of change, the problem of land and spatial policy', in J. Carney, R. Hudson, and J. Lewis (eds.), *Regions in Crisis* (London: Croom Helm).
Lovering, J. (1990), 'Fordism's unknown successor: A comment on Scott's theory of flexible accumulation and the re-emergence of regional economies', *International Journal of Urban and Regional Research*, 14: 159–74.
McGrath, M. E., and Hoole, R. W. (1992), 'Manufacturing's new economies of scale', *Harvard Business Review* (May–June), 94–102.
Martinez, J. I., and Jarillo, J. C. (1989), 'The evolution of research on coordination mechanisms in multinational corporations', *Journal of International Business Studies*, 20: 489–514.
Massey, D. (1979), 'In what sense a regional problem?', *Regional Studies*, 13: 233–43.
—— (1984), *Spatial Divisions of Labour: Social Structures and the Geography of Production* (London: Macmillan).
Morris, D., and Hergert, M. (1987), 'Trends in international collaborative agreements', *Columbia Journal of World Business*, 22: 15–21.
Moulaert, F., and Swyngedouw, E. (1989), 'Survey 15: A regulation approach to the geography of flexible production systems', *Environment and Planning D: Society and Space*, 7: 327–45.
Ohmae, K. (1985), *Triad Power: The Coming Shape of Global Competition* (New York: The Free Press).
—— (1990), *The Borderless World: Power and Strategy in the Interlinked Economy* (London: Collins).
Perlmutter, H. V. (1969), 'The tortuous evolution of the multinational corporation', *Columbia Journal of World Business* (Jan.–Feb.).
Piore, M., and Sabel, C. F. (1984), *The Second Industrial Divide* (New York: Basic Books).

Porter, M. E. (1986) (ed.), *Competition in Global Industries* (Boston: Harvard Business School Press).

—— (1990), *The Competitive Advantage of Nations* (London: Macmillan).

Reich, R. (1991), *The Work of Nations* (New York: Random House).

Sayer, A. (1985), 'Industry and space: A sympathetic critique of radical research', *Environment and Planning D: Society and Space*, 3: 3–29.

Schoenberger, E. (1991), 'Globalization and regionalization: New problems of time, distance and control in the multinational firm', Paper presented to the Annual Meeting of the Association of American Geographers, Miami.

Scott, A. J. (1988), *New Industrial Spaces: Flexible Production, Organisation and Regional Development in North America and Western Europe* (London: Pion).

Stalk, G. R. (1988), 'Time—the next source of competitive advantage', *Harvard Business Review* (July–Aug.), 41–51.

—— and Hout, T. (1990), *Competing against Time: How Time-Based Competition is Reshaping Global Markets* (New York: The Free Press).

Storper, M. (1992), 'The limits to globalization: Technology districts and international trade', *Economic Geography*, 68: 60–93.

—— and Scott, A. J. (1989), 'The geographical foundations and social regulation of flexible production complexes', in J. Wolch and M. Dear (eds.), *The Power of Geography: How Territory Shapes Social Life* (Boston: Unwin Hyman).

Thompson, J. D. (1967), *Organisations in Action* (New York: McGraw Hill).

Tickell, A. T., and Peck, J. A. (1992), 'Accumulation, regulation and the geographies of post-Fordism: Missing links in regulationist research', *Progress in Human Geography*, 16: 190–218.

Walker, R. (1989), 'A requiem for corporate geography: New directions in industrial organisation, the production of place and uneven development', *Geografiska Annaler*, 71B: 43–68.

Wells, P. E., and Cooke, P. N. (1991), 'The geography of international strategic alliances', *Environment and Planning A*, 23: 87–106.

Zukin, S., and DiMaggio, P. (1990) (eds.), *Structures of Capital* (Cambridge: Cambridge University Press).

3

Global Agro-Food Complexes and the Refashioning of Rural Europe

Sarah Whatmore

> Agriculture is a fundamental element in European society. Europe is built on a balance between the town and the countryside. No one will push Europe into sacrificing its agriculture. We are not going to depopulate rural Europe in order to hand the world's agricultural market to the Americans.
>
> Jacques Delors, President of the European Commission, Speech to the Houston Summit, 1990. (Quoted in Forsell 1991: 16.)

INTRODUCTION

The industrialization of agriculture in the post-war period has undermined its image as the mainstay of a traditional rural world in advanced industrial countries. At the same time, counterurbanization and a rural shift in secondary and tertiary employment against a steady decline in primary livelihoods has produced a dramatic, if uneven, reshaping of rural social and economic relations centred on the consumption of rurality as aesthetic and environmental capital for tourism, residence, and recreation (Cloke and Goodwin 1992). This dissociation of rural and agricultural space has produced a conceptual hiatus in research with the category rural repeatedly being declared theoretically redundant (Marsden *et al.* 1990). Critical research which has sought to engage with the theoretical projects of political economy and regulation theory has become increasingly bifurcated, with agriculture being reconceived in terms of a globally integrated industrial food complex (Friedland 1991), while those concerned with the restructuring of rural society have largely redefined their focus as the 'non-agricultural fractions [of capital]' (Cloke 1989: 179).

The erosion of rural distinctiveness projected by these theoretical discourses has secured a passage for those marginalized by disciplinary conventions as rural or agricultural specialists into the 'mainstream' of social science research with many fruitful consequences (Buttel *et al.* 1990). But as Raymond Williams has observed, the 'rural economy has never been solely an agricultural economy . . . only as a consequence of the Industrial revolution [has] the idea (though never the full practice) of rural economy and society been limited to agriculture' (1989: 231). Equally important is the obverse of this argument. Agriculture has never been restricted in its economic or social significance to a rural domain but has consistently forged a basic link between consumption and production, environment and society which has fashioned rural *and* urban relations simultaneously (Goodman and Redclift 1991). As Mormont reminds us, the oppositional designations rural and urban are socially constructed and formative in the 'search for identity and the organization of common cause' (1990: 41) including, as the opening quotation illustrates, the fabrication of national and European statehood.

The bifurcation of rural research, most marked in Britain, is in danger of losing sight of the continuing significance of agriculture as a definitive land-use in shaping rural relations and identities and of the spatial unevenness of its transformation into a globalized industrial complex. Interpretations of globalization as a homogenizing process are particularly inappropriate in a European context, in light of the diversity of food consumption and production practices, and the salience of rurality as an oppositional standard in social and political struggles over the identity, autonomy, and survival of those living on the land. Thus, without seeking to diminish the significance of non-agricultural influences on the restructuring of rural Europe, or to restrict that of agriculture to a 'rural' domain, this chapter is primarily concerned with reconsidering the spatial contours of the globalization of the agro-food complex in terms of its embeddedness in local social and environmental change in rural regions.

Like much of the literature on globalization, that concerned with the restructuring of the agro-food complex frequently conflates *global* processes, i.e. socio-economic processes understood as being articulated at a world scale, with *general* processes, i.e. socio-economic processes which are understood as being spatially unfixed and universal. Such an interpretation of globalization is liable to (at least) two criticisms. The first concerns the eradication of social agency in such accounts and recalls Lipietz's warning against the twin errors of deducing changes on the ground from immanent laws and of reifying analytical concepts such that they acquire an internal rationality that appears to orchestrate the pattern and experience of change in all places (1986: 17). Recognized as a label for the social construction of an integrated network of agencies, institutions, and values, whether in the form of the global accumulation strategies of multinational corporations or the global reach of media and telecommunications technologies, our attention is shifted to the social interactions (including their hermeneutic dimension) through which globalization is substantively and symbolically constituted.

The second criticism concerns the failure to adequately theorize the spatiality of globalization as a socially contested, rather than logical, process. Building on the themes explored in the 'new regional geography'

(Sayer 1989; Thrift 1990), globalization can be seen to assume and *reproduce* spatially uneven social and economic relations in the sense that the agents and institutions pursuing global economic strategies confront a historically differentiated landscape and, configured with other institutionalized interests and strategies, fashion new spatial divisions in their wake. Perhaps more importantly, such accumulation strategies and interests do not proceed unchecked but are realized through discursive and practical struggles over the identity and resources of particular places with other interests for whom the locality, region, or nation may be a more significant terrain and object of strategic action (Pred and Watts 1992).

In combination, these points suggest that in acknowledging a process of globalization in the structure of accumulation in the late capitalist world, it is dangerous, analytically and politically, to project the institutions and production filieres at its leading edge as all encompassing. The dynamics of social cohesion, human reproduction, political representation, and cultural identification are spatially embedded in different, often contradictory, forms to those of capital accumulation; impeding and enabling the process of globalization in important ways that cannot be reduced to a function of 'capital'. Equally, where most attention has been given to the 'new industrial spaces' or 'hot spots' of global capitalism (Tickell and Peck 1992) there is much to be learnt about the process of globalization from the 'cold spots' in this late twentieth-century economic landscape; places which may represent sites of active resistance, whether in defence of traditional or progressive alternatives, rather than the passive marginalization that is frequently implied. Above all, it seems imperative to reconnect the global with the local through the institutional networks, including those of capital, which mediate them at national and regional levels (Giddens 1990).

The first part of this chapter recounts the established story of the globalization of the agro-food complex as it has been developed in the agricultural sociology literature in a regulationist account that has followed parallel, but largely independent, lines to the debates in economic geography about 'post-Fordist' industrial restructuring (Gertler 1992). As will be seen, this story centres on the displacement of agriculture and rural land-use from the technological and competitive dynamics of accumulation and concentrates on the structural relations of multinational corporate capital and supranational regulatory institutions which have forged a global regime of food production and consumption. The second part of the paper attempts to reconsider the globalization process as a contested and locally embedded process by looking at its differential impact on the social and environmental relations of agriculture, the institutional bases of farmers' strategies for survival on the land, and their implications for the spatial reconfiguration of rural Europe.

THE GLOBALIZATION OF THE AGRO-FOOD COMPLEX: A REGULATIONIST TALE

Globalization is central to the development of what has been called the '*new* political economy of agriculture', a term relating both to the changing organizational structure of capitalist agriculture in the post-war period and to a refocusing of research questions and analyses. As Friedland succinctly puts it:

the present situation is one in which the connotations of 'farming'—in particular, rurality and community, but also other categories that are limited to national economies, nation-states, and national societies—are giving way to vertically and horizontally integrated production, processing and distribution of generic inputs for mass marketable foodstuffs (1991: 3–4).

The distinguishing feature of the 'new' political economy research agenda is the rejection of the established analytical separation of agriculture from wider processes of industrial restructuring as a 'special' or 'backward case', and the reappraisal of the strategic significance of food production and consumption to the formation of the metropolitan nation state (McMichael and Myhre 1991) and the rise of archetypal Fordist and 'post-Fordist' regional economies (Page and Walker 1991; Paloscia 1991).

This approach has taken the analysis of agriculture beyond the farm-gate to look at the social, economic, and technological interrelations between the activities of food raising (i.e. farming as a rural land-use), science and the agro-technology industries, food processing and retailing, and state institutions at national and supranational levels which regulate agricultural production and trade, and food consumption. It had its origins in the United States with studies of specific 'commodity chains', typically of highly industrialized agricultural commodities like salad crops (Friedland 1984). More recently, it has been taken up in the analysis of 'agri-food complexes', a term coined by Friedmann (1982) to describe the industrial processes and relations of the production and consumption of particular food products, and 'food regimes' (after Aglietta) linking international relations of food production and consumption to forms of accumulation broadly distinguishing periods of capitalist transformation (Friedmann 1982; Friedmann and McMichael 1989). While much effort has been put into linking the dynamics of the restructuring of the agro-food system as a production filiere to wider transformations in the structure of capitalist accumulation and social regulation, for the purposes of this chapter it is useful to outline the central arguments of each analytical theme in turn.

The Technological Transformation of a Biological
Production Process

Goodman and Redclift (1991: 90) argue that, 'to understand accumulation in the agri-food system it is essential to recognise that industrialization has taken a markedly different course from other production systems'. This distinctiveness is centred on the biological foundations of agricultural production in which the growth and reproduction of plant and animal life that are its products are bound to a series of biological processes and cycles which have resisted direct and uniform transformation by capitalist relations of production (Goodman *et al.* 1987). The objective limits of these biological constraints on accumulation are not fixed but they have made agriculture as a land-based activity unattractive to the direct involvement of industrial capital and its characteristic business forms and divisions of labour. Instead, capitalist restructuring has centred on reducing the dependency of agri-food accumulation on land through the technological modification of biological processes and the valorization of agricultural products off the land in the manufacture of agro-technologies and various processing, packaging, and preserving processes which add value to food products manufactured from farm output.

These 'non-farm' sectors of the agro-food system have become highly industrialized. in the post-war period and characterized by oligopolistic production and market structures dominated by large transnational corporations (Le Heron 1988). According to Goodman *et al.* (1987), the rapid growth of these sectors has taken place through two discontinuous but persistent processes: 'appropriationism'—in which elements once integral to the agricultural production process are extracted and transformed into industrial activities and then reincorporated into agriculture as inputs, and 'substitutionism'—in which agricultural products are reduced first to an industrial input and increasingly replaced by manufactured or synthetic non-agricultural components (Goodman *et al.* 1987: 2).

The development of biotechnologies in the agro-food sector holds the potential to further revolutionize these processes, enabling genetic transfer across species barriers as 'inputs' into agricultural production and the transformation of its products into generic inputs, as say starch or protein sources, in the manufacture of industrial foodstuffs. In this way, such technologies threaten 'the relative exclusiveness of different eco-systems' and so, some have argued, represent radical possibilities for the spatial reorganization of the agro-food system (Goodman and Wilkinson 1990: 39). The effects of these technological processes are illustrated in Fig. 3.1, using the example of the reconstitution of the production and consumption of the potato.

Agriculture is differentially integrated into this agro-food system in

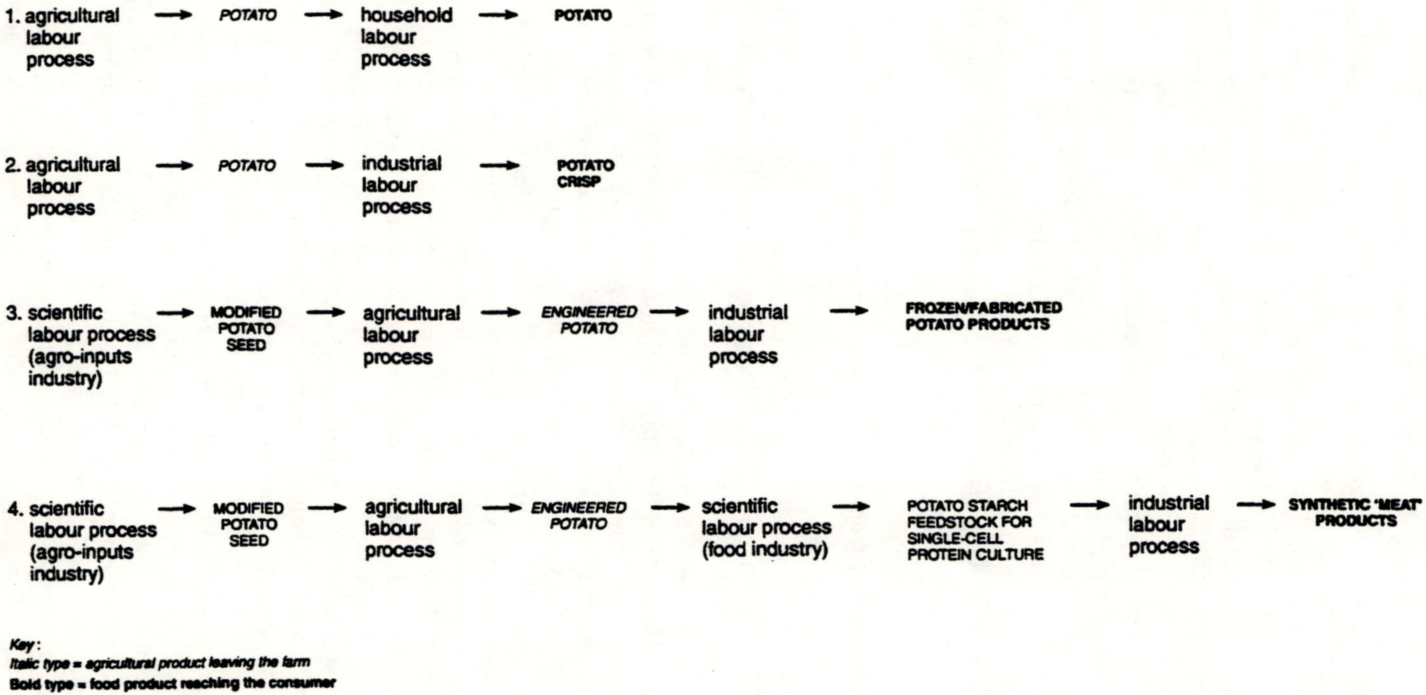

Fig. 3.1. *The industrial transformation of the potato*

1. agricultural → *POTATO* → household → **POTATO**
   labour   labour
   process   process

2. agricultural → *POTATO* → industrial → **POTATO**
   labour   labour   **CRISP**
   process   process

3. scientific → *MODIFIED* → agricultural → *ENGINEERED* → industrial → **FROZEN/FABRICATED**
   labour process   *POTATO*   labour   *POTATO*   labour   **POTATO PRODUCTS**
   (agro-inputs   *SEED*   process   process
   industry)

4. scientific → *MODIFIED* → agricultural → *ENGINEERED* → scientific → **POTATO STARCH** → industrial → **SYNTHETIC 'MEAT'**
   labour process   *POTATO*   labour   *POTATO*   labour process **FEEDSTOCK FOR**   labour   **PRODUCTS**
   (agro-inputs   *SEED*   process   (food industry) **SINGLE-CELL**   process
   industry)   **PROTEIN CULTURE**

Key:
Italic type = agricultural product leaving the farm
Bold type = food product reaching the consumer

terms of the sectoral characteristics of different types of agricultural commodity and the social relations of different forms of agricultural production; divisions which are spatially embedded and significant in ways that I shall return to later. These processes of technological innovation and diffusion, and the social divisions and economic relations of the agro-food complex associated with them, have been inextricably tied to state policies and agencies in the form of public research institutions and extension services, technological subsidies and agricultural price supports (Busch *et al.* 1989), and to the expanded role of banking capital and credit relations in agriculture (Le Heron 1991; Marsden and Whatmore 1994). The role of this 'agricultural technology/policy model' (Goodman and Redclift 1990: 24) in the post-war consolidation of a 'Fordist' regime of accumulation in the United States, its globalization and enduring crisis is the focus of the second analytical theme identified above.

Globalization of Capital–State Relations and the International Farm Crisis

The regulation of the production and consumption relations of food has been argued by Friedmann (1982), Friedmann and McMichael (1989), and McMichael and Myhre (1991) to play a pivotal role in the wider development of regimes of accumulation and their associated modes of social regulation because of the strategic significance of food in constituting the wage relation. Friedmann, who pioneered work in this field, has identified two international food regimes both of which are linked to the hegemonic rise of the United States in the world economy from the second half of the nineteenth century until the mid-1970s. The first food regime centred on the productive advantages of settler agriculture and the extensive production of wheat and cattle in North America and Australia in the period 1870–1914 and is implicated in the decline of Europe's colonial dominion. The rapid expansion of the market availability of the products of this 'grain-feed-livestock' complex through lower priced 'new world' imports marks the beginnings of an industrial diet in the non-colonial world based on animal protein consumption.

The interwar years saw this food regime in crisis as its productivity gains outstripped market capacity and national governments, led by the New Deal programmes in the United States, established permanent regulatory institutions to support agricultural product prices and take responsibility for storing and disposing of surpluses (Gilbert and Howe 1991). The second international food regime, dated to the period 1945–73, brought the consolidation of this agro-industrial model in the United States, centred on a corporatist alliance between agricultural science and policy-making institutions, farmers' organizations, and agro-food corporations, and its extension through US programmes of post-war reconstruction, such as

Marshall aid in Europe. The standardization of an industrial, or 'Fordist' diet (Kenney *et al*. 1991) was further established through state-supported advances in agro-food technologies which displaced traditional and seasonal foodstuffs with mass-produced foods, or what Friedmann has called a 'durable foods complex'.

In the 1960s and 1970s this US-centred complex expanded further through the extension of technological and food aid packages to Third World countries. As a result, much of the food-production capacity of developing countries has been switched from staple food crops to 'exotic' or unseasonal export crops to the First World, or feed crops for intensive livestock production, while turning them into net importers of foodstuffs from the United States and other Western states. However, as Tubiana (1989) has pointed out, the diffusion of the technological and institutional relations of this second food regime to other advanced industrial countries, notably the EEC and the Cairns Group, has undermined the hegemony of the United States in the international agro-food system. Moreover, the escalating fiscal costs of financing this post-war 'technology–policy model' of agricultural development have destabilized the transnational institutional foundations of the 'Fordist' mode of regulating food production and consumption (McMichael and Myhre 1991).

This destabilization, or crisis, in what Peterson (1990) has called 'productivist agriculture', has a number of unsynchronized but overlapping dimensions which have reasserted the significance of the biological foundations of agro-food production in unforeseen ways. The first dimension is a crisis of *production* which centres on the problems of surplus production and indebtedness amongst agricultural producers faced with escalating input costs and declining product prices. Second is a crisis of *regulation*, resulting from the growing political and institutional tensions in the corporatist agricultural policy–technology model between the regulation of global agro-food trade, on the one hand, and of regional and national farm support, on the other; a tension most recently illustrated in the dilemma of the EC negotiators in the current GATT round. A third dimension is a crisis of *legitimation*, which centres on the politicization of consumer concerns about the consequences of industrialized agriculture for food safety and the farmed environment at national and local levels (Sauer 1990; Flynn *et al*. 1991).

Coincident with this crisis in the Fordist institutions of regulation in advanced capitalist agriculture has been a growing dissatisfaction with the analytical terms in which the above research accounts of the globalization of the agro-food system are couched. In the tradition of much Marxist agrarian political economy, the regulationist concept of 'food regime' imposes a categorical logic on the restructuring of the production and consumption of food, representing it as a coherent process determined by the structural requirements of capital accumulation (Munton 1992: 32). In

consequence, the *differential* integration of agriculture within the technological and institutional relations of global agro-food complexes, alluded to
earlier, disappears from view and with it the significance of social agency
and local specificity to the constitution of such complexes.

As Table 3.1 indicates for the UK case, the social and economic relations
of agriculture are markedly more diffuse than those of other sectors in the
agro-food system.

In contrast to efforts in other branches of industrial geography to
establish similarly categorical accounts of a 'post-Fordist' regime of
accumulation, agricultural sociologists and geographers have increasingly
sought to make sense of the social heterogeneity of agriculture which lies
behind its distinctive profile in the agro-food system. This has been
encouraged by the fact that many of the features claimed to be distinctive
to post-Fordism as a 'flexible' regime of accumulation are long-established
characteristics of the social relations of agriculture. Most notably, simple
or domestic forms of commodity production, centred on the family-based
organization of land, capital, and labour continue to dominate Western
European agriculture and to articulate with transnational capital through a
range of locally negotiated contractual arrangements and nationally
embedded state, and state-supported, institutions (see e.g. Munton and
Marsden 1991; Long 1989; van der Ploeg 1990). This picture is further
complicated by the changing social relations of agricultural consumption
and the increasingly significant role of agricultural producers in the
political economy and cultural encoding not just of food, but of the
commoditization of the farmed environment as a landscape for leisure.

From this perspective, global agro-food complexes are represented as
socially, rather than logically, constituted and globalization is treated as a
contested process in which the disorganizing impact of the global strategies
and networks of transnational agro-food capital is seen to condition (rather
than determine), and to be conditioned by, those of agricultural producers
and consumers. Such a perspective is developed with respect to the
refashioning of rural Europe in the next section.

REFASHIONING RURAL EUROPE:
MAKING SPACE FOR AGENCY

In broad terms, those pursuing an account of globalization which opens up
'analytical' space for social agency and local diversity are shifting attention
to the interconnectivity between social, cultural, and political institutions
and relations of production *and consumption*, rather crudely assigned in
regulationist theory to the 'mode of social regulation' and about which, as
Tickell and Peck (1992) point out, little work has been done. In the agro-
food context, much of this attention has focused on the growing

TABLE 3.1. *The structure of the UK food chain in 1987*

Sector	Total sales (£ m 000)	Concentration
Agro-inputs:		
machinery	1.0	4 companies/ 77% market
animal feedstuffs	2.7	4 companies/ 57% market
fertilizers	0.8	3 companies/ 90% market
agrochemicals	0.3	5 companies/ 65% market
SUBTOTAL	4.8	
Agriculture	11.4	144,100 full-time holdings/ 97.4% output
Food manufacturing	26.4	10 companies/ 44% market
Food processing	32.3	4 companies/ 47% market

Source: After Ward (1990: 441).

significance of environmentalism in agricultural policy-making and, more broadly, in securing political authority in the advanced industrial world (Buttel 1992; Lowe 1992). Such work highlights the significance of new alliances and diversified struggles over rural resources and identities involving agricultural producers, local agro-food capitals, food consumers, environmentalists, and consumers of rurality. These alliances and struggles are articulated with the institutions of the local and national state apparatus and must be negotiated by transnational agro-food corporations (Almås 1990; Sauer 1990).

In the case of Western Europe, the point raised at the end of the last section about the social heterogeneity and political organization of agricultural producers requires elaboration. Despite its absolute and relative decline in terms of traditional economic indicators of rural production or employment, the social and economic relations of agriculture continue to dominate rural land-use and property rights and to represent cultural and environmental authenticity in rural life in policy and popular discourses (Marsden and Murdoch 1991). With the exception of Belgium (31%), France (47%), and Luxembourg (52%), at least two-thirds of agricultural land in all EC countries is owned by the farm operator. With the exception of the United Kingdom, less than 20% of farms are larger than 50 ha, while more than 74% of farms in southern European countries are less than 10 ha in size (see Table 3.2).

TABLE 3.2. *Distribution of farm size by EC country, 1985*
(all figures in rounded %ages)

Country	< 10ha		10–50ha		> 50ha	
	N	A	N	A	N	A
Belgium	46	12	49	64	5	24
Denmark	19	5	66	52	15	43
Eire	32	7	59	59	9	34
France	31	5	52	46	17	49
Germany[a]	48	12	47	63	5	25
Greece	89	54	6	30	5	16
Italy	84	34	14	34	2	33
Netherlands	43	12	53	70	4	18
Portugal	91	24	8	16	2	55
Spain	74	17	21	28	6	55
UK	25	2	42	16	34	83

Key: N = % of holdings.
A = % of farmed area.
[a] Figure for West Germany, prior to reunification.
Source: Commission of the European Community (1988).

Unlike in parts of the United States (Fitzsimmons 1986) corporate agro-food capital, including some of Europe's largest companies (e.g. Carrefour (French), Unilever (Anglo-Dutch), and Delta (Greek)), has made limited *direct* inroads into the farm labour process in Europe (Pugliese 1991) or into agricultural property ownership (Whatmore *et al.* 1990). Instead their influence has been exerted indirectly, through a complex of 'external relations' (Marsden *et al.* 1987) or 'interface networks' (Hawkins 1991) with agricultural capital–labour in the form of production contracts for the supply of technological inputs and associated technical assistance, and for the sale of agricultural products to food processors and retailers.

The organization of agricultural capital, land, and labour remains primarily in 'family' hands, centred on patriarchal kinship and household relations (Whatmore 1991). This distinctive industrial structure sustains very diverse and locally sedimented cultural and economic practices which articulate with transnational capital in complex, and often contradictory, ways in such key areas as knowledge and technology transfer, product and market specialization, and the institutional embeddedness of farming interests in the machinery of agricultural regulation across Europe (Hamilton 1985). This institutional embeddedness has augmented their political weight in local and national arenas far in excess of farmers' numerical or economic strength, and underpins a cultural significance which defies regulationist accounts.

In this context, the globalization of production and consumption relations in the agro-food sector is constituted through the contested interface between transnational corporate capital and institutionalized farming interests; an interface intricately bound into the fabric of the state apparatus and political culture at all scales from the local to the supranational (see e.g. Cox *et al.* 1987; Neville-Rolfe 1985). The distinctive social profile of agriculture noted above makes this process of contestation highly differentiated in its institutional and spatial configuration.

This is well illustrated by studies of the contrasting institutional structures and political cultures of agricultural regulation in Norway (Almås 1990) and Sweden (Peterson 1990). Norway retains the most heavily subsidized agricultural sector in Europe, and a domestic food market still largely outside the global reach of transnational agro-food corporations. Socially dominated by family-owned and -worked farms, the political weight of farming interests is firmly institutionalized through two union organizations and a series of agricultural commodity cooperatives, both with traditionally strong ties to the Social Democratic party which has dominated post-war government. The incorporation of farming organizations within this centre-left political culture in Norway is argued by Almås to have played a key role in Norway's resistance to membership of the EEC and the deregulation of agricultural production and the liberalization of agro-food trade which it would herald. The election of a Conservative government in Sweden in the late 1980s, by contrast, saw the marginalization of the institutional channels of farmers' political power and has resulted in the most dramatic programme of agricultural deregulation yet seen in Europe.

More generally, the unevenness of this contested restructuring of agro-food production and consumption is spatially embedded, simultaneously shaping, as well as being shaped by, locally and regionally specific agricultural practices, institutions, and alliances (Marsden *et al.* 1987). In the terms of the 'food regime' story of globalization, 80% of Europe's agricultural production derives from just 20% of agricultural businesses concentrated in agro-industrial 'hot spots' such as the Paris Basin (wheat), East Anglia (grain), Emilia-Romagna (milk), and the Netherlands (hydroponic crops) (Byé and Fonte 1991). But, as Map 3.1 illustrates, the areas 'marginal' to global agro-food accumulation and left out of this story represent much of rural Europe and many of its agricultural products, particularly in the south. Such 'marginal' areas are classified as 'lagging behind in development' in EC terminology, meaning that agricultural production is less market-dependent and less integrated into the technological and institutional relations of industrialized agro-food complexes. As van der Ploeg (1992*a*) points out, such an interpretation only makes sense within a unilinear deterministic model of industrial restructuring in which production relations at the leading edge of the model constitute a template

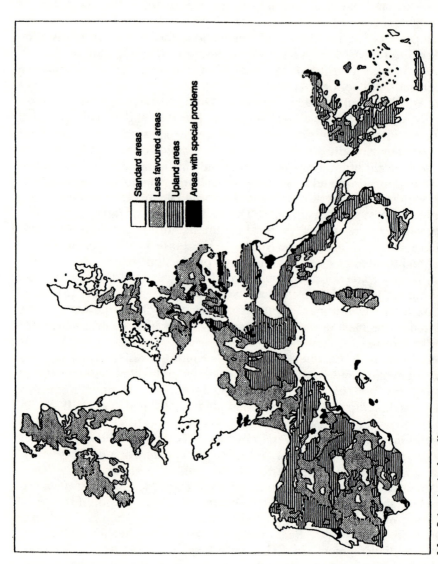

MAP 3.1. *Agriculturally marginal areas in the European Community*
Source: based on 1986 designations under EC directive 75/268

for the rest. He argues instead that differentiation and heterogeneity are inherent characteristics of capitalist agriculture in which leading-edge market and technological relations do not determine how farming is carried out but provide the context in which different positions are actively negotiated.

Deconstructing this category of 'marginal' farm businesses or regions van der Ploeg suggests a more variegated process of technology transfer and market integration in the restructuring of agriculture, identifying two alternative 'positions' to the orthodox interpretation of 'marginality' as farms or areas 'lagging behind' the leading edge. The first, which he calls 'vanguard farming', are areas which endeavour to transfer the technological and market relations of the 'hot spots' to 'marginalized areas' themselves. This relies upon increased dependence on outside elements (technologies, contract relations, and credit) and intervention (subsidies and technical assistance) to 'modernize' local farm structures; a strategy of 'exogenous growth' (van der Ploeg 1992*b*: 5) embodied in EC regional development policy. The second, which he calls 'alternative farming', are areas which endeavour to resist technology transfer and market integration and to retain non-commoditized systems of reproduction and artisanal practices and knowledges. This depends on institutionalizing traditional knowledge and techniques, through the informal structures of the local economy and labour-market, and their artisanal and ecological products through market networks and policy support systems beyond the local; a strategy of 'endogenous development' (van der Ploeg 1992*b*: 8).

Both strategies involve the articulation of local with global strategies and institutions, but on very different terms. Where the first is defined by the interests, and embodies the perspective, of transnational corporate capital, the second seeks to subvert the position of marginality by reinvesting in local knowledge, ecology, and culture and securing their viability through the creation of new, or the reworking of established, institutional networks beyond the locality. While there is a real danger of romanticizing such 'strategies of resistance', this rethinking of the 'cold spots' in the reconfiguration of rural Europe begins to address the experience of agricultural restructuring in many regions, particularly in the south (Hajimichalis and Vaiou 1987); an experience effectively silenced by the categorical logic of the 'global food regime'.

Numerous examples of such 'endogenous strategies' have been documented recently in Europe, including cooperative technological and marketing organizations, alternative farming unions, and their involvement in the regional and national policy-making process (de Haan and van der Ploeg 1992). But, perhaps the most important point to emerge from these cases for the success of such strategies in practice is the importance of political alliances between farming and other organized interest groups. Such alliances link farmers in one direction with broader local production interests in food processing and distribution which accounts for an estimated 30% of non-agricultural rural employment in southern Europe, with some 90% of agro-food firms across Europe employing fewer than twenty people (Traill 1988). Of even greater significance, alliances with organized consumer and environmental interests around the issues of food quality and safety and ecology are beginning to challenge the hegemony of the industrial, or 'productivist', policy consensus which has dominated the agricultural regulatory regime in the post-war period.

First, as Flynn *et al.* (1991) point out, the accumulative strategies of global agro-food corporations are dependent on the food consumption practices of people in particular national, regional, and cultural contexts. In Europe, the regulation of these practices with respect to the admission of products onto the market is still firmly wedded to national institutions and protocols which have become an increasingly politicized focus of opposition from consumer groups (Flynn *et al.* 1991: 165). The cultural embeddedness of food consumption tastes in Europe has already been recognized by the European Commission as a major obstacle to the realization of the level of market integration for foodstuffs achieved in the United States. In this context, consumer resistance to industrialized food products, particularly those associated with new biotechnologies, represents an important basis for new political alliances and marketing networks involving farming groups practising ecological or artisanal agricultural production methods (Clunies-Ross and Hildyard 1992).

Second, the environmental consequences, both ecological and aesthetic, of industrial agriculture have become another important focus for the politicization of agriculture (Sauer 1990). While it is farmers who have borne the immediate costs of new environmental regulations, for example over water pollution, the agro-technology industries have also been faced with global overcapacity in traditional bulk inputs sectors, such as nitrogenous fertilizers and animal feeds, and new research and development costs in launching new 'cleaner' technologies. The significance of this environmental challenge has been augmented by the commoditization of rural landscapes in the leisure, tourist, and residential development industries. Where landscape used to be a byproduct of farming practices, and other rural land-uses, it is increasingly becoming a product in its own right, involving a reorientation of capital and state agencies towards producing and reproducing landscapes valued by consumers of the countryside (Potter 1992).

The spatial reconfiguration of rural Europe resulting from the contested integration of agricultural commodities, livelihoods, and landscapes within global agro-food complexes centres, then, on a process of differentiation, not homogenization. Leading-edge technological and market relations associated with such complexes are realized unevenly, concentrated in rural regions characterized by highly industrialized agricultural practices and capitalized farm structures. Much of rural Europe, particularly southern Europe, is 'marginal' to these complexes, not in the sense of being untouched by the global strategies of agro-food corporations but of negotiating and institutionalizing alternative agricultural strategies and networks, increasingly bound up with the newly commoditized consumption of rurality. Fig. 3.2 illustrates this process of spatial differentiation schematically.

Following van der Ploeg (1992*b*) the status of marginality is deconstructed and a spectrum of four ideal types of local rural strategy identified, each involving a different reconfiguration of agricultural and non-agricultural capital and land-uses, and of the policy regimes characterizing national and supranational state regulation. From this perspective, the politics of agricultural restructuring in Europe can be argued to be being 'relocalized' and the livelihoods of those living on the land drawn into much wider struggles over the cultural and material resources of rurality. Here, globalization of the agro-food system is constituted in sectorally, socially, and spatially differentiated ways in which transnational corporate agro-food capital is variably, rather than uniformly, articulated through specific regional alliances, institutions, and regulatory regimes (Fitzsimmons 1986).

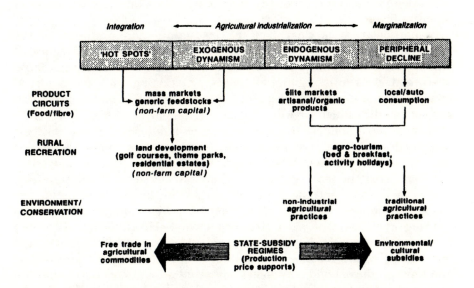

FIG. 3.2. *Agricultural restructuring and the fragmentation of rural space*

CONCLUSIONS

Reversing the lens of regulation theory from a preoccupation with the 'hot spots' of late capitalist accumulation to focus instead on the 'arid wastes' of the periphery, this chapter has sought to suggest an alternative account of globalization as a socially contested, rather than logical, process. This account is premised on the conviction that without some engagement with theoretical debates about social agency and more detailed empirical work which attempts to learn from the diverse experiences of those making a living outside, as well as within, the heartland of corporate capital, critical economic research is in danger of reproducing, rather than challenging, dominant power relations as the daily struggles of those on the margins become not so much a matter for pessimism, but of irrelevance in our accounts of the restructuring process.

The reworking of the concept of globalization sketched here has centred on the development of an intermediary level of analysis, concerned with the institutional fabric, public and private, through which the social relations and strategic interests of transnational capital and those of other producer and consumer organizations and alliances are interwoven in particular times and places. As the case of the agro-food system highlights, state agencies and policies at different spatial scales are, above all, political arenas in which the global strategies of corporate capital, however powerfully articulated, are contested by counter-interests more, or less, well-organized and institutionally embedded within the state apparatus at different levels.

In particular, it is clear that the institutions and protocols of the regulatory regime are an important site of struggle for the mobilization of alternative 'development' strategies, in the case of agriculture, those concerned with establishing alternative marketing and technological relations which reconnect local and regional production and consumption of food and the farmed environment. As Friedmann and McMichael (1989: 113) observe, ultimately the success of such strategies depends on their combination and co-ordination at higher levels of state policy-making and they are rightly cautious about overestimating their effectiveness. But the fiscal crisis in the regulatory regime associated with the agro-food system in Europe, and the associated decline in the power of the nation state, provide some new grounds for optimism. More importantly, there is growing evidence that such struggles have been effectively extended beyond the strategic interests of competing fractions of capital and increasingly centre on new alliances between petty capitalist agricultural and food producers and consumers mobilized around concerns about the impact of the industrialization on food and environmental quality.

In rethinking the process of globalization with respect to the agro-food system, this chapter has drawn attention to some distinctive features of this 'production filiere' and their implications for the (re)construction of rurality. Where farming, or, more specifically, food production, once cemented the association between agriculture and rural areas, the globalization of the agro-food system has displaced much of the agricultural labour process from the land and recast the terms on which the social and economic relations of agriculture articulate with the cultural politics and environmental resources of rurality. The *differential* integration of agriculture, socially, sectorally, and regionally, into a globalized regime of food production and consumption can be seen to represent both the product, and the possibility, of struggles over the industrialization of food, the degradation of the environment, and the defence of cultural identity in which those living on the land play a strategic, but increasingly diversified, role within much wider social and political realignments across Europe. These same struggles should also lend new theoretical weight to a conception of rurality as a dynamic configuration of socio-environmental relations with powerful symbolic and material effects.

NOTE

I should like to thank Nigel Thrift and Ash Amin for bearing with me through the numerous delays in getting this written; the organizers of the PURE programme for the opportunity to be involved; and Terry Marsden and Peter Maskell for stimulating conversations about potatoes.

REFERENCES

Almås, R. (1990), 'Globalization of food trade and its impact on reshaping a national food policy system', paper presented to the XII World Congress of Sociology, Madrid, July.

Busch, L., Bonnano, A., and Lacy W. (1989), 'Science, technology and the restructuring of agriculture', *Sociologia Ruralis*, 29/2: 118–30.

Buttel, F. (1992), 'Environmentalization: Origins, processes and implications for rural social change', *Rural Sociology*, 57/1: 1–27.

—— Larsen, O., and Gillespie, G. (1990), *The Sociology of Agriculture* (Westport, Conn.: Greenwood Press).

Byé, P., and Fonte, M. (1991), 'Technical change in agriculture and new functions for rural spaces in Europe', paper presented at the American Sociological Association, Cincinnati, Ohio.

Cloke, P. (1989), 'Rural geography and political economy', in R. Peet and N. Thrift (eds.), *New Models in Geography* (London: Unwin Hyman).

—— and Goodwin, M. (1992), 'Conceptualising countryside change: From post-Fordism to structured coherence', *Transactions of the Institute of British Geographers*, 17/3: 321–36.

Clunies-Ross, T., and Hildyard, N. (1992), *The Politics of Industrial Agriculture* (London: Earthscan).

Commission of the European Community (1988), *The Situation of Agriculture in Europe* (Brussels).

Cox, G., Lowe, P., and Winter, M. (1987), 'Farmers and the state: A crisis for corporatism', *Political Quarterly*, 58/1: 73–81.

Fitzsimmons, M. (1986), 'The new industrial agriculture: The regional integration of speciality crop production', *Economic Geography*, 62/4: 334–53.

Flynn, A., Marsden, T., and Ward, N. (1991), 'Managing food? A critical perspective on the British experience', in INRA, *Changement technique et restructuration de l'industrie agro-alimentaire en Europe* (Grenoble: INRA).

Forsell, L. (1991), 'Recent trends in EC and GATT. Consequences for rural Europe', in R. Almås and N. With (eds.), *Rural Futures in an International World* (Trondheim: Centre for Rural Research).

Friedland, W. (1984), 'Commodity systems analysis: An approach to the sociology of agriculture', *Research in Rural Sociology and Development*, 1: 221–35.

—— (1991), 'Shaping the new political economy of advanced capitalist agriculture', introd. to id., L. Busch, F. Buttel, and A. Rudy (eds.), *Towards a New Political Economy of Agriculture* (Boulder, Colo.: Westview Press).

Friedmann, H. (1982), 'The political economy of food: The rise and fall of the postwar international food order', *American Journal of Sociology*, 88 (suppl.): 248–86.

—— (1988), 'The family farm and international food regimes', in T. Shanin (ed.), *Peasants and Peasant Societies* (2nd edn., Harmondsworth: Penguin).

—— and McMichael P. (1989), 'Agriculture and the state system: The rise and decline of national agricultures, 1870 to the present', *Sociologia Ruralis*, 29: 93–117.

Gertler, M. (1992), 'Flexibility revisited: Districts, nation-states and the forces of production', *Transactions of the Institute of British Geographers*, 17/3: 259–78.

Giddens, A. (1990), *The Consequences of Modernity* (Cambridge: Polity).

Gilbert, J., and Howe, C. (1991), 'Beyond "state vs. society": Theories of the state and new deal agricultural policies', *American Sociological Review*, 56: 204–20.

Goodman, D., Sorj, A., and Wilkinson, J. (1987), *From Farming to Biotechnology* (Oxford: Basil Blackwell).

—— and Redclift, M. (1990) (eds.), 'The farm crisis and the food system: Some reflections on a new agenda', in T. Marsden and J. Little (eds.), *Political, Social and Economic Perspectives of the International Food System* (Aldershot: Avebury).

—— —— (1991), *Refashioning Nature: Food, Ecology and Culture* (London: Routledge).

—— and Wilkinson, J. (1990), 'Patterns of research and innovation in the modern agro-food system', in P. Lowe, T. Marsden, and S. Whatmore (eds.), *Technological Change and the Rural Environment* (London: David Fulton Publishers).

de Haan, H., and van der Ploeg, J. D. (1992) (eds.), *Endogenous Regional Development in Europe: Theory, Method and Practice* (Brussels: European Commission DGVI).

Hadjimichalis, C., and Vaiou, D. (1987), 'Changing patterns of uneven regional development and forms of social reproduction in Greece', *Environment and Planning D, Society and Space*, 5: 319–33.

Hamilton, P. (1985), 'Small farmers and food production in Western Europe', *International Social Science Journal*, 3: 345–60.

Hawkins, E. (1991), 'Changing technologies: Negotiating autonomy on Cheshire farms', Ph.D. thesis (Southbank Polytechnic, London).

Kenney, M., Lobao, L., Curry, J., and Goe, R. (1991), 'Agriculture in US Fordism: The integration of the productive consumer', in Friedland *et al.* (eds.), *Towards a New Political Economy of Agriculture*.

Le Heron, R. (1988), 'Food and fibre production under capitalism: A conceptual agenda', *Progress in Human Geography*, 12/3: 409–30.

—— (1991), 'New Zealand agriculture and changes in the agriculture-finance relation during the 1980s', *Environment and Planning A*, 23: 1653–70.

Lipietz, A. (1986), 'New tendencies in the international division of labor: Regimes of accumulation and modes of regulation', in A. Scott and M. Storper (eds.), *Production, Work and Territory* (Winchester, Mass.: Allen and Unwin).

Long, N. (1989) (ed.), *Encounters at the Interface: A Perspective on Social Discontinuities in Rural Development* (Wageningen Studies in Sociology, 27: Wageningen: Agricultural University).

Lowe, P. (1992) (ed.), 'Industrial agriculture and environmental regulation', special issue of *Sociologica Ruralis*, 32/1.

McMichael, P., and D. Myhre (1991), 'Global regulation versus the nation-state: Agro-food systems and the new politics of capital', *Capital and Class*, 43: 83–105.

Marsden, T., Whatmore, S., and Munton, R. (1987), 'Uneven development and the restructuring process in British agriculture: A preliminary exploration', *Journal of Rural Studies*, 3/4: 297–308.

—— Lowe, P., and Whatmore, S. (1990), introd. to eid. (eds.), *Rural Restructuring: Global Processes and Local Responses* (London: David Fulton Publishers).

—— and Murdoch, J. (1991), 'Restructuring rurality: Key areas for development in assessing rural change' (Working paper 4: Countryside change initiative. University of Newcastle upon Tyne).

—— and Whatmore, S. (1994), 'Finance capital and food system restructuring: Global dynamics and their national incorporation', in P. McMichael (ed.), *Agro-food System Restructuring in the Late Twentieth Century: Comparative and Global Perspectives* (Ithaca, NY: Cornell University Press).

Molinero, F. (1990), *Los espacios rurales* (Barcelona: Ariel Geograffía).

Mormont, M. (1990), 'Who is rural, or how to be rural? Towards a sociology of the rural', in T. Marsden *et al.* (eds.), *Rural Restructuring*, 21–44.

Munton, R. (1992), 'The uneven development of capitalist agriculture: The repositioning of agriculture within the food system', in K. Hoggart (ed.),

Agricultural Change, Environment and Economy: Essays in Honour of W. B. Morgan (London: Mansell), 25–48.

——— and Marsden, T. (1991), 'Dualism or diversity in family farming?', *Geoforum*, 22/1: 105–17.

Neville-Rolfe, E. (1985), *The Politics of Agriculture in the European Community* (London: European Centre for Political Studies).

Page, B., and Walker, R. (1991), 'From settlement to Fordism: The agro-industrial revolution in the American midwest', *Economic Geography*, 67/4: 281–315.

Paloscia, R. (1991), 'Agriculture and diffused manufacturing in the Terza Italia', in S. Whatmore, P. Lowe, and T. Marsden (eds.), *Rural Enterprise: Shifting Perspectives on Small-scale Production* (London: David Fulton Publishers).

Peterson, M. (1990), 'Paradigmatic shift in agriculture: Global effects and the Swedish Response', in T. Marsden *et al.* (eds.), *Rural Restructuring*.

van der Ploeg, J. D. (1990), *Labour, Markets and Agricultural Production* (Boulder, Colo.: Westview Press).

——— (1992a), 'The reconstitution of locality: Technology and labour in modern agriculture', in T. Marsden, P. Lowe, and S. Whatmore (eds.), *Labour and Locality: Uneven Development and the Rural Labour Process* (London: David Fulton Publishers).

——— (1992b), 'Styles of farming: An introductory note on concepts and methodology', in de Haan and van der Ploeg (1992).

Potter, C. (1992), 'Approaching the limits: Farming contraction and environmental conservation in the UK', in D. Goodman and M. Redclift (eds.), *The International Farm Crisis* (London: Macmillan).

Pred, A., and Watts, M. (1992), *Reworking Modernity* (Rutgers: Rutgers University Press).

Pugliese, E. (1991), 'Agriculture and the new division of labour', in W. Friedland *et al.* (eds.), *Towards a New Political Economy of Agriculture*.

Sauer, M. (1990), 'Fordist modernisation of German agriculture and the future of family farms', *Sociologia Ruralis*, 25/3–4: 260–79.

Sayer, A. (1989), 'The "new" regional geography and problems of narrative', *Environment and Planning D, Society and Space*, 7: 253–76.

Thrift, N. (1990), 'For a new regional geography 1', *Progress in Human Geography*, 14/2: 272–80.

Tickell, A., and Peck, J. (1992), 'Accumulation, regulation and the geographies of post-Fordism: Missing links in regulationist research', *Progress in Human Geography*, 16/2: 190–218.

Traill, B. (1988), 'Technology and food: Aims and findings of the EC Fast Programme's research into the prospects and needs of the European food system', mimeo, DG XII-FAST.

Tubiana, L. (1989), 'World trade in agricultural products: From global regulation to market fragmentation', in D. Goodman and M. Redclift (eds.), *The International Farm Crisis* (London: Macmillan).

Ward, N. (1990), 'A preliminary analysis of the UK food chain', *Food Policy* (Oct.), 439–41.

Whatmore, S. (1991), *Farming Women: Gender, Work and Family Enterprise* (London: Macmillan).

——— Munton, R., and Marsden, T. (1990), 'The rural restructuring process:

Emerging divisions of agricultural property rights', *Regional Studies*, 24/3: 235–45.

Williams, R. (1989), 'Between the country and the city', repr. in *Resources of Hope* (London: Verso).

4

The Uneven Landscape of Innovation Poles: Local Embeddedness and Global Networks

Franz Tödtling

INTRODUCTION

Recent changes in industrialized economies and their spatial transformation seem to be of a fundamental and a long-term nature. Some scholars argue that Fordism is transformed into post-Fordism, others observe the emergence of a new long wave of development. As a consequence the spatial system is seen to change dramatically as has been indicated in a 'break down or reversal of longstanding core periphery structures' (Aydalot 1984), the 'emergence of new sunbelt regions' and of 'Silicon Landscapes' (Hall and Markusen 1985) as well as of 'new industrial districts and spaces' (Piore and Sabel 1984; Scott 1988). These changes are said to affect not just production but also higher-level functions such as R&D, innovation, and decision-making. In what follows it will be investigated to what extent a move from Fordism towards post-Fordism is changing in particular the landscape of innovation poles in Europe.

The production and innovation process under Fordism was conceived to take place along a strict division of labour—to a large extent inside large firms. This was also reflected in space; the largest agglomerations tended to function as centres of decision-making, R&D, and product innovation while the peripheral regions were locations of standardized production under external control. Thus, the fortune of localities and regions depended strongly on their respective role within this division of labour.

The 'crisis' of Fordism and an emerging post-Fordism seem to change this structure of production, decision-making, and innovation considerably (Piore and Sabel 1984; Kern and Schumann 1985; Scott 1988; Storper and Walker 1989; Harvey 1990). There is a shift expected from large-scale standardized production towards small-scale customized production, and from the large hierarchically structured firm towards more decentralized firms, as well as to networks of firms. In the innovation process there is a much larger role for incremental change, and for small firms as well as networks of firms.

Together, these developments seem to lead to a stronger role for 'place' in economic development and in the innovation process. The strongest expression of this is the debate about the role of the 'local milieu' in innovation (Aydalot 1986; Aydalot and Keeble 1988; Maillat 1991). It is also argued that the traditional spatial hierarchy of the innovation process might be replaced by a more dispersed pattern or even reversed.

In the following sections I want to analyse the changing landscape of innovation poles in the European context. I will start with the changing nature of the innovation process going along with a move from Fordism towards post-Fordism. Then the embedding into the local 'milieu' (the role of local institutions and relations) and of global networks in the innovation process will be discussed. Finally I will present evidence concerning the existing and emerging landscape of innovation poles in Europe.

It will be shown that an emerging post-Fordism is bringing about substantial changes in this respect. However, the nature of this change is the result of a coexistence of traditional and new organizational models in space, rather than a basic reversal of past structures. In fact, it seems that despite some development in new areas, post-Fordism tends to strengthen existing innovation poles, namely large metropolitan areas as locations of technological innovation.

THE CHANGING NATURE OF THE INNOVATION PROCESS MOVING FROM FORDISM TOWARDS POST-FORDISM

The innovation process characteristic of Fordism has been shaped by basic principles such as

- a high share of standardized products for large markets (mass production);
- a dominant role of the large corporation organizing production and innovation;
- a highly developed division of labour and a clear-cut separation of conception and execution inside firms, whereby this 'technical' division of labour is also reflected in space (Westaway 1974; Watts 1981; Massey 1984).

The introduction of new products, their change towards mass products, and the related firm-behaviour has been described by product-cycle theory (Norton and Rees 1979; Utterback 1979). Firms develop new products via their R&D departments, launching them on the market and subsequently reaping the benefits of growing demand and sales. In the course of maturation the product becomes more standardized, competition increases, and it becomes necessary to cut costs. New, more standardized production processes are introduced and scale economies are realized. The division of labour, particularly within firms, increases, external delivery links become more stable, and they are extended over large spatial scales.

This process has a distinctive spatial pattern, particularly since the large firm, which is the motor of this process, is able to specialize locationally and draws benefits from a spatial division of labour (Bade 1984; Massey 1984): R&D facilities are optimally located in large agglomerations near universities and highly trained labour. Also, the new products are best produced here since the markets as well as a variety of suppliers are close. In the course of maturation, when cost considerations get more important and standardization occurs, production moves towards cheap labour in rural areas, very often supported by regional policy (Erickson and Leinbach 1979). In the late stage of the cycle it may become necessary to move to still cheaper locations such as the southern periphery of Europe or to newly industrializing countries (Fröbel *et al.* 1977). There is a spatial hierarchy of innovation implied: R&D and product innovation are concentrated in the largest agglomerations of the highly developed countries, while process change or a lack of innovation are observed in locations of the late stage such as old industrialized areas and the rural periphery (Tödtling 1990).

The emerging post-Fordist accumulation and regulation regime (see Moulaert *et al.* 1988; Harvey 1990; Benko and Dunford 1991; Cooke and Morgan 1991), in contrast, is characterized by

- a diversification of consumer demand and, consequently, a lower standardization of products (customization);
- use of flexible technologies, organizations, and labour practices;
- a certain decentralization of functions within large firms (bringing some of the higher-level functions back to the production level) and a bias towards horizontal instead of vertical information flows; and
- a more prominent role of small firms partly through vertical disintegration of large firms, spin offs, and subcontracting relations to large firms.
- Finally there is an increasing importance of institutions not only at the local and regional level but also at newly emerging international levels (such as the EC) as actors in economic development.

There are considerable implications for the innovation process. Due to the frequent changes of products and the shortening of product life cycles, technological change becomes more continuous (Nelson and Winter 1982; Dosi 1988). Furthermore, it is increasingly a non-linear process which need not necessarily start with technical inventions but may also begin at the production level or may be initiated by suppliers or customers (Hakanson 1987; von Hippel 1988). Furthermore, technological change is seen as proceeding in an evolutionary fashion along trajectories and is strongly shaped by existing routines. Such routines are partly rooted in the local and regional institutions and the milieu so that specific regional trajectories may emerge (Aydalot 1988; Malecki 1991).

The speed and cost of technological change increase and they lead to a reinforced division of labour in the innovation process. This occurs between large and small firms, universities, and public research institutions, as well as other public institutions (transfer agencies, regional and local development institutions). Thus, an increasing variety of actors and institutions are involved. Some of the firms engage in global networks, others establish links at the local and regional level (Cooke and Morgan 1991; DeBresson and Walker 1991; Freeman 1991).

Both Fordism and post-Fordism describe rather extreme ideal types of production and regulation which in reality cannot be found in a pure form. In particular, the transition to post-Fordism is strongly disputed. While some of the changes such as the increasing diversification and customization of products as well as the trend towards more flexible technology, organization, and labour relations have been broadly accepted, it has been questioned whether this constitutes a new production and regulation regime leading to a general return to 'place' (Amin and Robins 1990; Sayer 1989). In the light of a recent wave of mergers and acquisitions and the increasing rather than decreasing dominance of large firms (Martinelli and Schoenberger 1991), the picture of an economy based on small and flexible firms, concentrated in certain districts, has been put into question.

LOCAL EMBEDDEDNESS AND GLOBAL NETWORKS IN THE INNOVATION PROCESS

The evolutionary character of technological change and the increasing social division of labour emphasize the relevance of both the local environment of firms (embeddedness in local networks and institutions) as well as of large-scale and global networks. Both, however, fulfil different functions and they have different importance for small as against large firms.

Local embeddedness has been stressed particularly by the 'milieu approach' (Aydalot 1986; Aydalot and Keeble 1988; Maillat 1991). The 'milieu' can be defined as the socio-economic environment of an area resulting from the interaction of firms, institutions, and labour. It is expected to lead to a common way of perceiving economic and technical problems and finding respective solutions (Maillat 1991). The milieu mainly fulfils the task of informal knowledge transfer through mobile labour, information links, supplier and customer links, as well as through cooperations at the local and regional level. Although there is evidence for the engagement of both large and small firms, the local milieu is said to be most important for small firms, since these are less able to maintain costly boundary-spanning functions (such as R&D) or to engage in large-scale networking (see Maillat 1991; Camagni 1991). The milieu is particularly considered to favour learning processes in the innovation process such as learning by doing and learning by using and interacting. An important element in the local milieu is education and training institutions, as well as institutions engaged in firm-formation and technology transfer such as incubation and innovation centres (Herrigel 1989; Tödtling and Tödtling-Schönhofer 1990; Cooke and Morgan 1991).

However, not just the embedding of firms into the local milieu can be observed, but also an increasing integration into international and global R&D networks (Castells 1989). This occurs both internally to firms through the location or acquisition of R&D units in several countries and in the form of interfirm arrangements (through various forms of R&D cooperation). Major forces of globalization of R&D are to be found in a changing macro-research environment (Howells 1990) such as the emergence of pervasive new technologies (information technologies, new materials, and biotechnologies) breaking down traditional barriers between technological disciplines. Then there is an increasing complexity of new technology to be observed leading to increasing costs and time-spans to develop new products while the life-span of new products on the other hand has been reduced. Firms are thus seeking to extend the market for new products as fast as possible, to build up a stronger research base through mergers and acquisitions, and/or to externalize certain steps of the R&D process.

The location of R&D activities of (multinational) firms in more than one country is, as Howells (1990) observes, not a totally recent phenomenon.[1] However, the occurrence as well as the nature of such international R&D activities has been changing since the mid-seventies. Large firms are increasingly locating R&D units in more than one country, some of them moving towards global R&D networks. Along with this, the nature of R&D activities is changing from mere transfer units adjusting products only to local markets, towards more autonomous R&D units which are involved in self-development of products and processes.

The second type of international and global R&D networks is of an interorganizational nature: firms, particularly the large ones, are linking into distant or global networks through carefully selected formal co-operations and strategic alliances (see Camagni 1991; Freeman 1991).[2] These allow them to organize complementary resources in the innovation process, reducing the development time of the new product, and/or opening up new markets more rapidly. Empirical work on this aspect has been undertaken more systematically by Hagedoorn and Schankenraad (1990) for selected technology areas (information technology, biotechnology, new materials). They show that these formal inter-firm networks have increased, particularly in the 1980s (90% of about 4,600 investigated cases), and that the vast majority of them (again 90%) occur between the most highly developed countries of the 'Triad' (Europe, United States, Japan). The major motives were the opening-up of new markets, the time-reduction of the development process, the specific technological competence of the partner, and the screening of technological opportunities. The long-run strategic positioning of the firm was in general a more important goal of the cooperations than the reduction of transaction costs (Hagedoorn and Schankenraad 1990).

International R&D networks are also increasingly supported by supranational institutions and programmes (Malecki 1991; Charles and Howells 1992). Within Europe there exist already a number of research and technology programmes aiming at the international cooperation of firms and research institutions.[3] It appears that it is mainly large firms which have been benefiting from these programmes. For SMEs it is rather difficult to be fast enough in getting the necessary information, organizing partner(s), and going through the application procedure.

Up to now the analysis has shown that there is an increasing relevance not just of local embeddedness (particularly for small firms) but also of international and global networks. From a few studies it appears that there are also relations between these different types of networks. International and global network links are sometimes established in order to tap into the knowledge base of a specific local or regional milieu. In the case of the Silicon Valley, for example, Gordon and Dilts (1988) have shown that a number of inward investments and cooperations from foreign firms to Silicon Valley firms had just this intention. On the other hand, Camagni (1991) has pointed to the fact that the local milieu also needs to be linked up to international and global networks in order to stay innovative and avoid decline (an 'entropic death') in the long run. Linkages between the local milieu and global networks may exist in various ways, for example through large firm–small firm cooperations and subcontracting, through spin-offs from large firms, or through mobile labour.

THE UNEVEN LANDSCAPE OF INNOVATION POLES—EMPIRICAL EVIDENCE

The changes in the organization of production and innovation described above and the increasing variety of actors and institutions involved at the local and regional as well as on the supranational levels lead to a more complex landscape of 'innovation poles' than has existed in the past. On the one hand, most of the large metropolitan regions remain important centres of innovation. On the other hand, new patterns of innovation are emerging. These are partly adjacent or near to the large metropolitan areas but partly also in newly industrializing intermediate or even peripheral locations.

Large Metropolitan Regions as Major Innovation Poles

Large metropolitan regions were already the major innovation poles in the Fordist era, concentrating headquarters as well as R&D functions of large firms. They still have superior locational conditions such as high levels of education, large numbers of research institutions, and highly qualified labour. They are also major nodes in the international communication and transport networks, so they tend to remain important poles of innovation. In fact, some of the post-Fordist conditions such as the increasing social division of labour or the high relevance of networks reinforce their position in that respect (Camagni 1988; Malecki 1991). Empirical evidence from quite a number of countries (Austria, West Germany, France, Italy, Switzerland, United Kingdom, United States)[4] shows the continuing dominance of large metropolitan regions in the innovation process. This evidence also partly reflects features of the traditional product cycle model:

- R&D activities are strongly concentrated in the largest agglomerations where a few large firms in certain sectors usually hold high shares of R&D employment and expenses.
- Product innovations are relatively more frequent in agglomerations or in highly developed regions than, for example, in old industrial or peripheral areas.
- On the other hand, there is clearly more emphasis on process innovation or a lack of innovation in old industrial areas and peripheral regions.

This pattern of innovation is additionally shaped by contingent factors in respective countries, such as the historical spatial and urban structure or the pattern of public policy. A more clear-cut spatial hierarchy of innovation seems to exist in countries with a concentrated spatial structure such as Austria, France, Italy, or the United Kingdom, countries with one or few dominating agglomerations. In countries with a more balanced spatial structure such as Germany, the Netherlands, or Switzerland the pattern is less clear-cut.

A very strong concentration of R&D activities has also been observed at the European level (both for countries and for level-II regions) showing the situation of a 'two-speed' community.[5] In addition, there is differentiation occurring between European cities (Drewett *et al.* 1992): some of them are better able than others to further improve their knowledge base or to attract high-level enterprise functions and activities (examples are London, Paris, Munich, and Milan). Other cities, for a variety of reasons, are losing their entrepreneurial and technological dynamics. To the latter group belong agglomerations dominated by old industries (e.g. older conurbations in the United Kingdom or the Ruhrgebiet in Germany) or agglomerations where bureaucratically organized firms and institutions have become dominant (Vienna). Around or adjacent to the most dynamic high-ranking agglomerations there are 'spillovers' to be observed. These cover not just population, manufacturing industries, and consumer services, but also advanced producer services as well as innovative firms. Examples of such innovation poles close to major agglomerations are the Science city south of Paris (Aydalot 1986; Scott 1988), or the high tech areas to the north and west of London (M4 Corridor, Cambridge: Segal 1985; Keeble 1988). These new growth areas have received a lot of attention in the literature as areas representing the 'geography of the 5th Kontratieff' (Hall and Markusen 1985). Their success was very often ascribed to the special role of universities and research parks functioning as nuclei of development. However, the development of these areas seems to be more related to the nearby agglomerations and their locational advantages than to indigenous factors.

New Spatial Patterns of Innovation?

Along with an emerging post-Fordism there are new organizational forms of innovation coming up which also partly modify the traditional spatial hierarchy. The major actors in these new models are public policy (the case of science parks and technology centres), large firms which are deconcentrating some of their R&D activities to subsidiaries and branches, and networks of firms. To what extent is a change in the geography of innovation involved and what is the role of place in these developments?

Science parks and technology centres as policy approaches

Science parks emerged in the late 1950s in the United States but grew rapidly and in large numbers particularly in the 1980s in the United States and Japan as well as in Europe. They can be found in quite different locations, both in dynamic and stagnating regions, in large metropolitan areas as well as in medium-sized cities.[6] However, there are considerable qualitative differences between these locations concerning R&D intensity and high tech orientation, employment skills, university links, as well as growth performance of science and technology parks.

For the United Kingdom, Massey *et al.* (1992) found a significant north–south divide with regard to many features of the eighty-three science parks existing in 1988. In the north they turned out to be younger, smaller, more likely to be based on small indigenous firms, and less 'scientific' than those in the dynamic south-east. Similarly, Sternberg (1988, 1990), investigating seventy technology and incubation centres founded in West Germany since 1983, has observed a higher 'technology-orientation' of centres in the more prosperous south of Germany as well as in the larger agglomerations as against those in the north or those in the assisted areas. Finally Luger and Goldstein (1991) in a comprehensive analysis of 116 US research parks have found that the more successful research parks, measured by employment growth, were more likely to be situated in sizeable and dynamic regions and they were usually of an older vintage and had good links to a first-class research university.[7] On the other hand, park failures, as indicated by closure or slow employment growth, were more often in smaller regions and in slower-growing counties and not related to a nearby research university.[8]

The role of links to local universities and to the local economy is evaluated differently. While both the US study (Luger and Goldstein 1991) and the German study (Sternberg 1988) consider the links to a local university as rather important for the success of the relevant park or centre, Massey *et al.* (1992) for the United Kingdom doubt their general importance. They found for the UK science parks that 'formal research links between academic institutions and establishments on science parks were no more evident than similar links with firms located off-park' (Massey *et al.* 1992: 38).

What have been the effects of these research and technology parks on regional development and on interregional disparities? The available evidence suggests that in many cases there are positive (direct and indirect) employment effects (Luger and Goldstein 1991), positive effects on firm-formation[9] as well as on the structure of firms and on skills of employment in the respective location. However, with a few exceptions, the overall impact on the respective regional economies remains limited, since the number of firms and jobs involved is usually not dramatically high.[10] Furthermore, it can be seen that these parks and centres are not bringing about substantial changes of the spatial pattern of innovation since the more sophisticated and more successful parks tend to be located either in large metropolitan or in more dynamic regions. Since they are more often supported by institutions at the local or regional level than by those on the federal level they just tend to reflect and even reinforce differences of economic strength between the regions rather than to reduce them.

R&D in the decentralized large firm

Peripheral regions very often have a high share of branch plants with standardized production and low levels of R&D and management functions. However, since markets are becoming increasingly volatile, the requirement of flexibility exerts a certain pressure to reintegrate production with higher-level functions such as R&D and marketing at the plant level. Does this bring about an upgrading of branch plants through the location of certain R&D and decision-making functions also in peripheral regions?

A well-known example of semi-autonomous branches in high tech industries is the UK Silicon Glen in Scotland, a peripheral economy which, due to the working of the Scottish Development Agency, was able to attract branch plants from a number of multinational companies (IBM,

NCR, Honeywell, DEC, Hewlett-Packard, Motorolla, NEC, Sun Micro-systems, Compaq). These partly had also more qualified high tech production and some R&D (Haug 1986). However, the overall effects on employment and skills have remained rather limited. Moreover, there were few links created to firms in the region and to universities and there were also few spin-offs. Being not strongly integrated into the regional economy, quite a number of plants have closed down or reduced employment in the early 1990s during a recession (Sutherland 1993).

Charles (1987), investigating the UK electronic industry, identifies the following mechanisms for the creation of more semi-autonomous units within large firms: the spin-off process (development of new entre-preneurial units in order to exploit new ideas and technologies), the acquisition of, and equity investments into, small firms to gain access to their technology as well as joint ventures. These new units remain semi-autonomous within the larger company and they have better employment characteristics and greater linkages to the local economy than the traditional branch plant. However, due to higher locational and skill requirements, they are in general primarily a core-region phenomenon and are less frequent in peripheral regions.

Other examples of semi-autonomous or upgraded peripheral branch plants can be found in various countries. A recent study on Austria has identified a number of subsidiaries and branch plants in 'modern' industries with certain R&D functions also in peripheral regions (Tödtling 1990). Although these upgraded plants have better skill effects and income effects in the region than 'pure' production branch plants, the reduction of interregional disparities (income or qualifications) is limited since the basic division of labour inside the large firms is not eliminated (Castells 1989; Martinelli and Schoenberger 1991).

Firm-networks and small-firm dynamics

Regions characterized by small-firm dynamics and networks have received a lot of attention in the literature (see e.g. Garofoli 1991; Cooke and Morgan 1991; Pyke and Sengenberger 1992). Although the model of the Third Italy (small-firm networks) has attracted most interest, the networks where at least one or a few large firms are involved seem to be more frequent (Storper and Harrison 1991). Networks are in fact complex phenomena and they may contrast strongly in their industrial and spatial organization (see Bergman *et al.* 1991; DeBresson and Walker 1991; Storper and Harrison 1991). In addition to the various actors involved (small and large firms, public institutions), they may have different governance structures: from a relatively egalitarian structure, to a loose hierarchy (coordinating firm), to a more rigid hierarchy (leading firm). From the spatial point of view, the main actors may be agglomerated or dispersed, maintaining local as well as long-distance links. Many of these firm-networks—like the research parks—cluster again around agglomerations or in industrialized areas, but many also include smaller cities in non-metropolitan locations (this is particularly the case in Italian districts). Storper and Harrison (1991) in addition show that regional economies may consist not just of one but of several types of firm-networks.

Large-firm–small-firm networks

This type of network is coordinated or dominated by large firms which externalize certain functions or production steps to small or intermediate-sized firms. Partly there are linkages involved at the regional level and the network takes an agglomerated form. Storper and Harrison (1991) give examples of agglomerated large-firm networks from such different areas as Silicon Valley (systems integrations and 'merchant' semiconductors in electronic industry) and Toulouse (aircraft industry) as well as from the Third Italy (Emilia-Romagna: machinery industry). Perrin (1988) and Cooke and Morgan (1991) bring examples and evidence from Baden Württemberg (car industry: Daimler-Benz, Porsche, Bosch, Audi; electronic industry: IBM, Hewlett-Packard, SEL, Sony), Hanover (Volkswagen), Milan (Fiat, Olivetti), the south of the Netherlands (Philips), and the south of France (Sophia Antipolis). Hansen (1990) identifies large-firm–small-firm networks in the Marseilles region (including the hinterland) as well as in the area of Montpellier (Languedoc-Roussillon).[11]

In many of these cases, the strength of the network relations at the regional level is assumed and not made explicit. In the case of Baden Württemberg Cooke and Morgan (1991: 26 f.) find that a large proportion of the suppliers can be found within the region of Baden Württemberg although 'these large firms are by no means averse to sourcing from outside the region or Germany itself'. Schmitz (1992: 100) in addition states that the purchases from outside Germany have been increasing at a faster rate.[12] Closer examination of other cases might well reveal that the spatial reach of these large-firm–small-firm networks is much larger than the respective region. This is certainly the case for many Austrian subcontractors which deliver mainly to the German car firms.[13]

The traditional subcontracting networks between large and small firms implied a rather clear division of labour in the production and innovation process. The large firms had the function of 'core-production' and technology development (product and process innovation) as well as marketing and distribution, the small firms contributed certain steps and components in the production process. Technological relations occurred only in the form of a faster diffusion of process technology towards the subcontractors and a certain pressure for the subcontractors to introduce new machinery to keep the required standard or to save costs. The question arises whether the more recent kinds of networks go beyond these traditional relations and include also R&D and product and process development in their division of labour. Up to now the answer to this question has not been very well documented. Cooke and Morgan (1991), studying the regions of Baden Württemberg, Emilia-Romagna, the Basque Country, and Wales, observe a move towards 'collaborative manufacturing' for reasons of enhancing and maintaining the quality of supplies and materials and of shortening the innovation process.[14] Grabher (1991) gives some evidence of such emerging 'innovation networks' for the German Ruhr area: large firms are restructuring from their traditional products (e.g. steel) towards more know-how and technology-intensive activities (e.g. engineering) and are utilizing the flexibility and innovative capacity of small firms by building up network-relations to them. Large firms and small firms fulfil different tasks and functions. The large firm as a general contractor organizes the whole project, providing finance and market access. Small firms develop, design, and produce special contributions within this network. This implies that, despite the stronger role of the small firms, the nature of the collaboration stays basically a hierarchical one (Storper and Harrison 1991).

Small-firm networks

Based on the experiences of a few 'success stories' such as high tech regions (Silicon Valley, Route 128, M4 Corridor), as well as some other dynamic regions (the Third Italy, Baden Württemberg), localized small-firm networks have been viewed as the clearest indication of a possible 'end of Fordism' (Piore and Sabel 1984). Instead of the large firms involved in mass production, localized networks of a number of small firms produce a variety of customized products in small batches, cooperating not just in production but also in distribution and in technological development (Garofoli 1991). This development has also been seen as evidence for a new importance of agglomeration economies and a concentration of such industries in 'new industrial spaces' (Scott 1988).

Competitive strategies of firms in such districts are clearly different from Fordist strategies since advantages are achieved by high quality, know-how, or design intensity as well as by flexibility instead of scale economies and low cost. Due to the customization of products there is frequent change of both products and processes (modifications, design changes, etc.), whereby the innovation process is also supported by the network. It helps to spread risks, combine resources and assets, and share know-how and experience.

These networks where innovative small firms are strongly involved are again to be found in various locations (see Aydalot 1986; Aydalot and Keeble 1988; Cooke and Morgan 1991; Pyke and Sengenberger 1992). Some of them, particularly the high tech industries, cluster around large metropolitan areas (such as London, Milan, Paris, Madrid, and Barcelona: see the section on large metropolitan regions, above), others are in intermediate regions with more diffused industry (the Third Italy: clothing, textiles, leather goods, ceramics, furniture; Baden Württemberg: machine-tool industry; West Jutland in Denmark: dressmaking and knitting, furniture). Finally, there are also examples in 'old industrialized' regions on their way to restructure (Swiss Jura: watch industry).

The constitution of firm-networks at the local and regional level is attributed not just to economic but also to socio-cultural factors resulting from the history of the region (Garofoli 1991). Together, these factors are said to constitute a specific local milieu conducive to firm-formation and -innovation (see the section on local embeddedness and global networks, above).[15] Aydalot and Keeble (1988) regard even innovations as a product of the local milieu by stating that 'it is often the local environment which is, in effect, the entrepreneur and innovator, rather than the firm'. Spatial proximity and territorial agglomeration are considered as important preconditions for the creation and the functioning of such an innovative milieu (Stöhr 1987; Camagni 1991). This is due to the fact that highly skilled personnel are not very mobile interregionally, that direct personal and informal contacts are highly relevant, and that synergy effects are expected from a common cultural, psychological, and political background.

An important element, which is part of the network and the milieu, are the public and semi-public institutions supporting the SMEs in such regions. These may be trade or business associations, chambers of industry and commerce, and technology transfer centres, as well as public programmes for innovation and firm-formation. These institutions provide, amongst other services, research capacity, contract research, training, technology transfer, business information, and finance for the small firms. Cooke and Morgan (1991: 43) consequently summarize the key elements of the 'networked region' as 'a thick layering of public and private industrial support institutions, high grade labor market intelligence and associated vocational training, rapid diffusion of technology transfer, a high degree of interfirm networking and, above all, receptive firms well disposed towards innovation'.

A particularly 'thick' tissue of such support institutions can be found in Baden Württemberg (Herrigel 1989; Schmitz 1992) as well as in the industrial districts of the Third Italy (Cooke and Morgan 1991; Brusco 1992). In both regions these institutions have quite a long history, casting doubt on the ready transferability of such institutions to other regions. It is difficult, furthermore, to evaluate the role and effectiveness of these institutions on the performance of firms in the regions and to draw direct causal links from one to the other. As Schmitz states 'there is a great deal of description of the policy but—with a few exceptions—there is little analysis of its effectiveness' (1992: 104).[16]

It has been suggested that localized small-firm networks might become a general model of innovation and regional development. However, there are both conceptual and empirical problems with this proposition. At the conceptual level, a clear-cut transition from Fordism towards post-Fordism and to a small-firm economy has been questioned on various grounds (Sayer 1989; Amin and Robins 1990). Although it is true that small firms have experienced a certain renaissance since the mid-seventies, the large firms are far from leaving the stage, as has been shown above (Martinelli and Schoenberger 1991). There is also no general return to 'place' and local embeddedness since, as we have shown above, large-scale and global networks as well as supranational institutions have at the same time increased their importance. Empirically, the evidence for localized small-firm networks becoming a general model for innovation and regional development has up to now also not been convincing:

- There are rather few regions in Europe and the United States which have been serving as standard examples of localized and dynamic small-firm networks ('success stories'). Other regions have also been cited (Zeitlin 1992) but the differences seem to be larger than the similarities.
- A closer look in fact reveals that these regions are very heterogeneous: they range from regions with family- and craft-based industries having sometimes very low levels of technology (many districts of the Third Italy)[17] to sophisticated high tech regions at the 'frontier' of technological development.[18]
- Neither the high tech regions nor the districts very often stay as 'small and beautiful' as they were at their early stage but are subject to rapid organizational change. On the one hand, in some districts small firms under competitive pressure are forming large groups in order to stay competitive. Some small firms are also growing and internationalizing their markets and networks. On the other hand, these regions also get penetrated and dominated by external large firms.[19]
- As was shown above, there are many more regions which are linked in one way or the other to large coordinating or leading firms, inside or outside the region (see above; as well as Storper and Harrison 1991) or which have no significant network links at all.
- The empirical evidence for the strength of linkages at the local and regional level and of local embeddedness is very often anecdotal rather than being systematically derived. There only exist case studies in which usually rather few firms and institutions have been interviewed. More systematic and representative studies in a broader comparative setting are lacking. Furthermore, in the existing studies there has often been an implicit bias in favour of the identification of such localized networks. All this makes broader generalization and evaluation of this phenomenon rather difficult.

CONCLUSIONS

From this chapter the following conclusions may be derived. First, the move towards post-Fordism is indeed bringing about substantial changes in the production and innovation process, its organization as well as its manifestation in space. The changes imply a more continuous and evolutionary character of innovation and a stronger need to integrate R&D, marketing, and production at the plant level. The linear innovation process, organized in a hierarchical way within firms, which has been typical for Fordism, is losing relevance. There is also more need to rely on external partners, both firms and public institutions. Consequently there are new organizational forms coming up such as research parks and innovation centres, a decentralization of R&D and innovation activities in large firms, and, last but not least, networks of innovators.

Second, along with these changes there are new patterns of innovation in space emerging both in and around metropolitan areas and also in non-metropolitan regions. However, these changes do not imply a basic reversal of past structures and trends. Instead, they constitute a more complex structure since past and new patterns coexist (Amin and Robins 1990). In fact, it seems that despite some development in new areas, the move towards post-Fordism tends to strengthen major metropolitan areas as innovation poles. Since these are still important nodes in national and international transport and communication networks as well as locations for high-level educational and research institutions, they also stay major locations for research parks and technology centres, for R&D functions of large firms, as well as for firm-networks.

Third, local and regional institutions and the 'embeddedness' of firms seem to have more relevance in the evolutionary innovation process. However, this cannot be interpreted as a general return to 'place' as an exclusive driving force for economic and technological development because of several reasons:

- Firms are becoming not just locally embedded but at the same time increasingly integrated into the global economy. Particularly in the 1980s international and global networks (both within and between large firms) as well as supranational institutions have increased their importance.
- Local embedding ('place') and international and global links have different relevance for different actors: SMEs clearly rely more on local networks and local institutions (with regard to information-gathering, markets, finance) than large firms and their branches. Large firms and their plants, in contrast, are very often quite independent from the local environment. However, this pattern does not always hold true since a considerable number of SMEs are also engaged in international and

global markets and networks and parts of large firms are also linked to the local environment.

- Local embedding and networks are dynamic and subject to change, since many firms are extending their markets and networks in the course of their maturation. On the other hand, external large firms are very often penetrating regional economies and tapping into local networks via takeovers or interfirm alliances.
- Finally we must point to the fact that the empirical evidence concerning the local embedding of firms and its role for the innovation process is still very limited. In contrast to some of the 'euphoric' literature, there seem to be relatively few cases of 'milieu-driven' innovative regions. Also the particular role and functioning of local networks and the local 'milieu' *vis-à-vis* global forces up to now has not been convincingly demonstrated.

Thus, from the analysis set out above, it appears that local firms and networks are becoming increasingly interlinked with global markets, with corporate hierarchies as well as networks. It is actually this interaction of global forces with specific local conditions and histories which is increasingly shaping the innovation process and consequently local and regional development.

NOTES

1. Until the 1960s the number of multinationals undertaking R&D abroad, according to Howells (1990), were rather few. By 1965 it was estimated for US multinationals that only 6.5% of their total R&D expenditure was undertaken abroad.
2. Freeman (1991: 502) defines formal inter-firm networks as joint ventures, research corporations, technology exchange agreements, minority holdings for technological reasons, licensing and second-sourcing agreements, subcontracting, research associations, government-sponsored joint R&D programmes, and computer networks for technical interchange.
3. At the European level such programmes are COST (since 1971), EUREKA (1985), and ESA (1987). Within the EC particularly the second (1987–91) and third (1990–4) Framework Programme for Research and Technology are supporting international R&D projects (Commission of the EC 1991).
4. Empirical studies by Tödtling (1990, 1992) (*Austria*); Planque (1983) (*France*); Meyer-Krahmer *et al.* (1984), Meyer-Krahmer (1985), Ewers and Fritsch (1987), Pfirrmann (1991) (*Germany*); Cappellin (1983) (*Italy*); Brugger (1985), Brugger and Stuckey (1987) (*Switzerland*); Thwaites (1982), Buswell (1983), Gillespie (1983), Howells (1984, 1988), Thwaites and Oakey (1985), Goddard,

Thwaites, and Gibbs (1986), Thwaites and Alderman (1988) (*UK*); Malecki (1980, 1983, 1991), Castells (1989) (*USA*).

5. In the mid-1980s more than 75% of R&D employment of the EC was in the three countries West Germany, France, and UK. On the other hand, the less-favoured member states (Italy, Ireland, Spain, Portugal, and Greece) representing 40% of the population had only 10% of the R&D employment in the community. The R&D gap of course is much larger at the regional level showing a very concentrated pattern of R&D in Europe (Commission of the EC 1987, 1988).

6. See Tödtling and Tödtling-Schönhofer (1990) for an overview; Sternberg (1988, 1990) and Dose and Drexler (1988) for West Germany; Luger and Goldstein (1991) for the USA; Massey *et al.* (1992) for the UK; Stöhr and Pönighaus (1992) for Japan.

7. More specifically Luger and Goldstein (1991: 175) state that regions are more likely to host successful research parks if they have (1) an existing base of R&D and high tech activity, (2) one or several research activities, medical schools, and/or engineering institutes, (3) good air services, (4) a well-developed network of infrastructure and business services, and (5) foresighted and effective business leaders. However, these locational characteristics are regarded neither as necessary nor sufficient to ensure park success.

8. Other examples of research parks and technology centres, which partly are also in non-metropolitan locations, can be found in France (technopoles such as Sophia Antipolis near Nice, or ZIRST in Grenoble), Spain (nine technology parks existing), Sweden (innovation centre in Chalmers in Göteborg, research park IDEON in Lund), and Finland (seven technology parks).

9. A closer look, however, reveals that it is rather the persistence of newly created firms which is enhanced than the formation rate (Sternberg 1988; Tödtling and Tödtling-Schönhofer 1990).

10. In the 116 US research parks there were in total about 150,000 employees (1989), in the 83 parks of the UK there were about 15,000 people employed (1990), in the 60 technology centres of West Germany there were somewhat more than 10,000 employees (1988). Although these numbers are not negligible, they are just a very small percentage of total employment in those countries and regions.

11. Compared to the artificially created and rather exclusive technopolis of Sophia Antipolis, Hansen (1990) finds a stronger impact of the networks on regional development in these latter cases, since small firms were more spontaneously involved.

12. We also have to keep in mind that the 'region' of Baden Württemberg is fairly large (having a population of 9.5 m); it is a larger economy than Austria.

13. There are in Austria about 230 firms with an employment of 135,000 engaged in subcontracting to the international car industry.

14. They state that in Baden Württemberg for firms such as Bosch 'design of components is now commonly a joint process in which an intrafirm network of the Purchasing, Engineering and Design departments works with two Preferred Supplier Status firms, only one of which will be awarded the production contract' (Cooke and Morgan 1991: 27).

15. The 'milieu' was, according to Maillat (1991), defined as the socio-economic

environment of an area resulting from the interaction of firms, institutions, and labour. It is expected to lead to a common way of perceiving economic and technical problems and finding respective solutions.

16. In fact for Baden Württemberg Schmitz (1992: 108) finds that 'It would be misleading to attribute the success of Baden Württemberg in the 1970's and 1980's solely to the neoconservative modernisation policy of Lothar Späth. The success preceded or coincided with his technology policy. At the same time it would be unreasonable to entirely discard claims that regional government helped industry to cope with challenges of the 1990's.' We also have to keep in mind that 'institutional thickness' is not just typical for dynamic districts, but can also be found in old industrial areas on their way to restructure such as the Ruhr area in Germany, the Swiss Jura, or the Obersteiermark in Austria (Tödtling and Tödtling-Schönhofer 1990).

17. Out of about sixty industrial districts in the Third Italy over fifty specialized in fashionware (textiles, clothing, shoes, and other leather goods) or wooden furniture (Amin and Robins 1990: 17).

18. Zeitlin (1992: 185) consequently states that 'In the face of these difficulties, it seems necessary to move away from a "thick", "closed" model of the industrial district based on a stylised account of a particular national experience towards a "thin", "open" model capable of generating a variety of empirically observable forms.'

19. In the case of Silicon Valley this has been shown by e.g. Gordon and Dilts (1988). For the Italian industrial districts Brusco (1992) has shown that there is a considerable change towards the formation of groups of firms as well as towards the internationalization of markets. The increasing role and dominance of large firms in such regions is by some defenders of the district model seen as a process of a 'double convergence' of large- and small-firm structures (Zeitlin 1992) 'as small firms in the districts build wider forms of common services often inspired by large-firm models, while large firms seek to recreate among their subsidiaries and subcontractors the collaborative relationships characteristic for small-firm districts' (Zeitlin 1992: 184).

REFERENCES

Amin, A., and Robins, K. (1990), 'The reemergence of regional economies? The mythical geography of flexible accumulation', *Environment and Planning D, Society and Space*, 8: 7–34.
—— Malmerg, A., and Maskell, P. (1991), 'Structural change and the geography of production in Europe', RURE working paper, WG1.
Aydalot, P. (1984), 'A la recherche des nouveaux dynamismes spatiaux', in id. (ed.), *Crise & Espace* (Paris: Economica).
—— (1986) (ed.), *Milieux innovateurs en Europe* (Paris: GREMI).
—— (1988), 'Technological trajectories and regional development in Europe', in id. and Keeble (1988).

—— and Keeble, D. (1988) (eds.), *High Technology Industry and Innovative Environments: The European Experience* (London: Routledge).

Bade, F. J. (1984), *Die funktionale Struktur der Wirtschaft und ihre räumliche Arbeitsteilung* (Berlin: Deutsches Institut für Wirtschaftsforschung (DIW)).

Benko, G., and Dunford, M. (1991), 'Structural change and the spatial organisation of the productive system: an introduction', in eid. (eds.), *Industrial Change and Regional Development: The Transformation of New Industrial Spaces* (London: Belhaven Press).

Bergman, E., Maier, G., and Tödtling, F. (1991) (eds.), *Regions Reconsidered: Economic Networks, Innovation and Local Development in Industrialised Countries* (London: Cassel).

Breheny, M. J., and McQuaid, R. W. (1987), *The Development of High Technology Industries: An International Survey* (London: Croom Helm).

Brugger, E. A. (1985), *Regionalwirtschaftliche Entwicklung—Strukturen, Akteure und Prozesse* (Bern: Hauptverlag).

—— and Stuckey, B. (1987), 'Regional economic structure and innovative behaviour in Switzerland', *Regional Studies*, 21: 241–54.

Brusco, S. (1992), 'Small firms and the provision of real services', in Pyke and Sengenberger (1992).

Buswell, R. J. (1983), 'Research and development and regional development: A review', in Gillespie (1983).

Camagni, R. (1988), 'Functional integration and locational shifts in new technology industry', in Aydalot and Keeble (1988).

—— (1991), 'Space, networks and technical change: An evolutionary approach', in id. (ed.), *Innovation Networks* (London: Belhaven Press).

Cappellin, R. (1983), 'Productivity growth and technological change in a regional perspective', *Giornale degli Economisti ed Annali di Economia*, 42: 459–82.

Castells, M. (1989), *The Informational City: Information Technology, Economic Restructuring and the Urban-Regional Process* (Cambridge, Mass.: Basil Blackwell).

Charles, D. R. (1987), 'Technical change and the decentralized corporation in the electronics industry: Regional policy implications', in K. Chapman and G. Humphrys (eds.), *Technical Change and Industrial Policy* (Oxford: Basil Blackwell).

—— and Howells, J. (1992), *Technology Transfer in Europe: Public and Private Networks* (London: Belhaven Press).

Commission of the European Communities (1987), Research and Technological Development in the Less Favoured Regions of the Community (STRIDE), Final report by J. Goddard, D. Charles, J. Howells, and A. Thwaites (Brussels).

—— (1988), Science and Technology for Regional Innovation and Development in Europe (STRIDE), Final Report by the National Board for Science and Technology (Brussels).

—— (1991), Forschungs- und Technologieförderung der EG. Ein Leitfaden für Antragsteller. Amt für amtliche Veröffentlichungen der EG (Brussels).

Cooke, P., and Morgan, K. (1991), 'The network paradigm: New departures in corporate & regional development' (Regional Industrial Research Report, 8; Cardiff).

DeBresson, C., and Walker, R. (1991) (eds.), Network of Innovators, special edn. of *Research Policy*, 20/5.

Dose, N., and Drexler, A. (1988) (eds.), *Technologieparks—Voraussetzungen, Bestandsaufnahme und Kritik* (Opladen: Westdeutscher Verlag).

Dosi, G. (1988), 'The nature of the innovative process', in id. *et al.* (eds.), *Technical Change and Economic Theory* (London: Pinter).

Drewett, R., Knight, R., and Schubert, U. (1992), The Future of European Cities: The Role of Science and Technology (Prospective Dossier, 4; FAST-MONITOR Programme, Commission of the European Communities).

Erickson, R. A., and Leinbach, T. R. (1979), 'Characteristics of branch plants attracted to nonmetropolitan areas', in R. E. Lonsdale and H. L. Seyler (eds.), *Nonmetropolitan Industrialization* (London: John Wiley).

Ewers, H. J., and Fritsch, M. (1987), Die räumliche Verbreitung von computergestützten Techniken in der Bundesrepublik Deutschland (Diskussionspapier, 120; Wirtschaftswissenschaftliche Dokumentation, TU Berlin).

Freeman, C. (1991), 'Networks of innovators: A synthesis of research issues', in DeBresson and Walker (1991), 499–514.

Fröbel, F., Heinrichs, J., and Kreye, O. (1977), *Die neue internationale Arbeitsteilung* (Reinbek bei Hamburg: Rororo).

Garofoli, G. (1991), 'Local networks, innovation and policy in Italian industrial districts', in Bergman, Maier, and Tödtling (1991).

Gillespie, A. (1983) (ed.), *Technological Change and Regional Development* (London Papers of Regional Science; London: Pion Ltd.).

Goddard, J., Thwaites, A., and Gibbs, D. (1986), 'The regional dimension to technological change in Great Britain', in A. Amin and J. B. Goddard (eds.), *Technological Change, Industrial Restructuring and Regional Development* (London: Allen and Unwin).

Gordon, R., and Dilts, A. (1988), 'High technology innovation and the global milieu: Small & medium sized enterprises in Silicon Valley', Paper prepared for Colloque GREMI II, Ascona, Apr.

Grabher, G. (1991), 'Rebuilding cathedrals in the desert: New patterns of cooperation between large and small firms in the coal, iron, and steel complex in the German Ruhr area', in Bergman, Maier, and Tödtling (1991).

Hagedoorn, J., and Schankenraad, J. (1990), 'Strategic partnering and technological cooperation', in B. Dankbaar, J. Groenewegen, and H. Schenk (eds.), *Perspectives in Industrial Organization* (Dordrecht: Kluwer).

Hakanson, H. (1987) (ed.), *Industrial Technological Development: A Network Approach* (London: Croom Helm).

Hall, P., and Markusen, A. (1985) (eds.), *Silicon Landscapes* (London: Allen and Unwin).

Hansen, N. (1990) 'Innovative regional milieux, small firms, and regional development: Evidence from Mediterranean France', *Annals of Regional Science*, 24: 107–23.

Harvey, D. (1990), *The Condition of Postmodernity: An Enquiry into the Origins of Cultural Change* (London: Basil Blackwell) (1st edn. 1989).

Haug, P. (1986), 'US high technology multinationals and Silicon Glen', *Regional Studies*, 20: 103–16.

Herrigel, G. B. (1989), 'The politics of large firm relations with industrial districts: A collision of organisational fields in Baden Württemberg', mimeo University of Chicago.

Howells, J. (1984), 'The location of Research and Development: Some observation and evidence from Britain', *Regional Studies*, 18: 13–29.

—— (1988), 'The location and organisation of Research and Development', Paper presented to the European Summer Institute of the RSA, Arco, July.

—— (1990), 'The internationalization of R&D and the development of global research networks', *Regional Studies*, 24: 495–512.

Keeble, D. (1988), 'High-technology industry and local environments in the United Kingdom', in Aydalot and Keeble (1988).

Kern, H., and Schumann, M. (1985), *Das Ende der Arbeitsteilung? Rationalisierung in der industriellen Produktion* (Munich: Beck Verlag) (1st edn. 1984).

Luger, M., and Goldstein, H. (1991), *Technology in the Garden: Research Parks & Regional Economic Development* (Chapel Hill, NC: University of North Carolina Press).

Maillat, D. (1991), 'The innovation process and the role of the milieu', in Bergman, Maier, and Tödtling (1991).

Malecki, E. J. (1980), 'Corporate organization of R&D and the location of technology activities', *Regional Studies*, 14: 219–34.

—— (1983), 'Technology and regional development: A survey', *APA Journal*, 50: 262–9.

—— (1988), 'Research and development and technology transfer in economic development: The role of regional technological capability', in R. Capellin and P. Nijkamp (eds.), *The Spatial Context of Technological Development* (Aldershot: Avebury).

—— (1991), *Technology and Economic Development: The Dynamics of Local, Regional and National Change* (Essex: Longman Scientific & Technical).

Markusen, A. (1987), *Profit Cycle, Oligopoly and Regional Development* (Cambridge, Mass.: MIT Press) (1st edn. 1985).

Martinelli, F., and Schoenberger, E. (1991), 'Oligopoly is alive and well: Notes for a broader discussion of flexible accumulation', in Benko and Dunford (eds.), *Industrial Change and Regional Development*.

Massey, D. (1984), *Spatial Divisions of Labor: Social Structures and the Geography of Production* (London: Macmillan).

—— Quintas, P., and Wield, D. (1992), *High Tech Fantasies: Science Parks in Society, Science and Space* (London: Routledge).

Meyer-Krahmer, F. (1985), 'Innovation behaviour and regional indigenous potential', *Regional Studies*, 19: 523–34.

—— Dittschar-Bischoff, R., Gudrum, U., and Kuntze, U. (1984), Erfassung regionaler Innovationsdefizite. Schriftenreihe des BM für Raumordnung, Bauwesen und Städtebau, 06.054, Bonn–Bad Godesberg.

Moulaert, F., Swyngedouw, E., and Wilson, P. (1988), 'Spatial responses to Fordist and post-Fordist accumulation and regulation', *Papers of the Regional Science Association*, 64: 11–23.

Müdespacher, A. (1987), 'Adoptionsverhalten der Schweizer Wirtschaft und regionale Aspekte der Diffusion der Neuerungen der Telematik', *Jahrbuch der Regionalwissenschaft*, 8: 106–34.

Nelson, R. R., and Winter, S. G. (1982), *An Evolutionary Theory of Economic Change* (Cambridge, Mass.: Harvard University Press).

Norton, R. D., and Rees, J. (1979), 'The product cycle and the spatial decentralisation of American manufacturing', *Regional Studies*, 13: 141–51.

Perrin, J. C. (1988), 'New technologies, local synergies and regional policies in Europe', in Aydalot and Keeble (1988).

Pfirrman, O. (1991), *Innovation und regionale Entwicklung—Eine empirische Analyse der Forschungs- Entwicklungs- und Innovationstätigkeit kleiner und mittlerer Unternehmen in der Bundesrepublik Deutschland 1978–1984* (Munich: Verlag Florentz).

Piore, M., and Sabel, J. (1984), *The Second Industrial Divide: Possibilities for Prosperity* (New York: Basic Books).

Planque, B. (1983), *Innovation et développement régional* (Paris: Economica).

Pyke, F., and Sengenberger, W. (1992) (eds.), *Industrial Districts and Local Economic Regeneration* (Geneva: International Institute for Labor Studies).

Sayer, A. (1989), 'Postfordism in question', *International Journal of Urban and Regional Research*, 13: 666–95.

Schmitz, H. (1992), 'Industrial districts: Model and reality in Baden-Württemberg, Germany', in Pyke and Sengenberger (1992).

Scott, A. J. (1987), 'Flexible production systems and regional development: The rise of new industrial spaces in North America and Western Europe', *International Journal of Urban and Regional Research*, 12/2: 171–85.

—— (1988), *New Industrial Spaces: Flexible Production Organization and Regional Development* (London: Pion).

Segal, N. (1985), 'The Cambridge Phenomenon', *Regional Studies*, 19/6: 563–70.

Sternberg, R. (1988), *Technologie- und Gründerzentren als Instrument kommunaler Wirtschaftförderung* (Dortmund).

—— (1990), 'Regionaler Informationstransfer—Die Rolle von Technologie und Gründerzentren in der Bundesrepublik Deutschland', in *Innovations- und Technologiezentren—Ein taugliches instrument der Regionalpolitik?* (ÖROK-Schriftenreihe, 81; Vienna).

Stöhr, W. (1987), 'Territorial innovation complexes', *Papers of the Regional Science Association*, 59: 29–44.

—— and Pönighaus, R. (1992), 'The effect of new technological and organizational infrastructure on urban and regional development: The case of the Japanese technopolis policy', *Regional Studies*, 26/7: 605–18.

Storper, M. (1986), 'Technology and new regional growth complexes, the economics of discontinuous spatial development', in P. Nijkamp (ed.), *Technological Change, Employment and Spatial Dynamics* (Heidelberg: Springer-Verlag).

—— and Walker, R. (1989), *The Capitalist Imperative: Territory, Technology and Industrial Growth* (Oxford: Basil Blackwell).

—— and Harrison, B. (1991), 'Flexibility, hierarchy and regional development: The changing structure of industrial production systems and their forms of governance in the 1990s', *Research Policy*, 20/5: 407–22.

Sutherland, E. (1993), *Silicon Glen—A technological Brigadoon? An Analysis of the Electronics and IT Industries in Scotland* (IIR-Discussion, 48; Institute for Urban and Regional Studies, University of Economics, Vienna).

Thwaites, A. (1982), 'Some evidence of regional variations in the introduction and diffusion of industrial products and processes within British manufacturing industry', *Regional Studies*, 16/5: 371–82.

—— and Oakey, R. (1985) (eds.), *The Regional Economic Impact of Technological Change* (London: Francis Pinter).

—— and Alderman, N. (1988), 'The location of R&D: Retrospect and prospect', in R. Cappellin and P. Nijkamp (eds.), *The Spatial Context of Technological Development* (Aldershot: Avebury).

Tödtling, F. (1990), *Räumliche Differenzierung betrieblicher Innovation— Erklärungsansätze und empirische Befunde für österreichische Regionen* (Berlin: Edition Sigma).

—— (1992), 'Technological change at the regional level: The role of location, firm structure and strategy', *Environment & Planning A*, 24: 1565–84.

—— and Tödtling-Schönhofer, H. (1990), *Innovations- und Technologietransferzentren als Instrumente einer regionalen Industriepolitik in Österreich* (ÖROK-Schriftenreihe, 81; Vienna).

Utterback, J. M. (1979), 'The dynamics of product and process innovation in industry', in C. Hill and J. M. Utterback (eds.), *Technological Innovation for a Dynamic Economy* (New York: Pergamon Press).

Von Hippel, E. (1988), *The Sources of Innovation* (Oxford: Oxford University Press).

Watts, H. D. (1981), *The Branch Plant Economy: A Study of External Control* (London: Longman).

Westaway, J. (1974), 'The spatial hierarchy of business organisations and its implications for the British urban system', *Regional Studies*, 8: 145–55.

Zeitlin, J. (1992), 'Industrial districts and local economics regeneration: Overview and comment', in Pyke and Sengenberger (1992).

5

Growth Regions under Duress: Renewal Strategies in Baden Württemberg and Emilia-Romagna

Philip Cooke and Kevin Morgan

INTRODUCTION

Experts in regional economic development have increasingly begun to turn away from the former panaceas of growth-pole theory, on the one hand, and trickle-down theory on the other. As Walter Stöhr (1990) has recently argued, the issues in question have been recognized as being far too complex to yield to such one-dimensional recipes. Others such as Aydalot (1986), sharing Stöhr's persuasion, have recognized that regional economic development is as much about non-economic phenomena as it is concerned with precisely economic variables. The wave of apparently marginal, even esoteric, research into the mechanisms associated with unorthodox developmental tendencies, built around small firms, industrial districts, or local productive systems has proven extraordinarily fruitful for academics and policy-analysts alike. Of key importance to both understanding and policy prescription is the recognition that contemporary regional economic success is inseparable from cultural, social, and institutional accomplishment.

Nowhere has this been more the case than in Baden Württemberg in Germany and Emilia-Romagna in Italy. These two European regions display, in different ways, many of the archetypes of the 'intelligent region' (Cooke and Morgan 1991). They are both well possessed of intermediary institutions between business and government which quickly assimilate, relay, and transmit information. They are fortunate enough to have governmental institutions and personnel that are keen to learn from others and to apply the lessons. Finally, they also have the social, cultural, and institutional wit to be intelligent about their own performance, to self-monitor, evaluate, and alter policies, intermediaries, and people; they are prepared, as it were, to change a winning team. This approach has served the people and businesses of these regions well, so long as the economies remain relatively self-contained. But the onset of the 1990s marks the emergence of a heightened globalization (Cooke *et al.* 1992) in economic affairs, and with it a serious challenge to and questioning of the efficacity of their endogenous industrial systems, supporting intermediary institutions and networks of business relationships.

This chapter explores these two regions from the standpoint of observers who wish to judge just how good these regional systems are at a point in history when they look as if they are in danger of being tested to destruction. This test stems from the interlinked forces of, on the one hand, the need to control monumentally high public deficits, with their associated inflationary tendencies, and, on the other, the erosion of intra-EC market barriers and the globalizing whirlwind of Japanese competition. In regional economies dependent upon small-firm domination of substantial sectors of industry, the state and other intermediaries are of key importance in assisting firms to afford important skills, technology, and marketing services which they would otherwise find uneconomic to purchase.

Thus, in what follows, the inherited models of development displayed in the two regions are each laid out; these are then placed in the context of the larger field in which they operate, with references made to key changes occurring at that level. Thereafter, the nature of the institutional response to the looming crises of the two regions is explored and conclusions are drawn. In writing about the two regions, we report a large amount of original research. But it would be churlish not to mention the importance to our understanding of the two regions of key contributions to the literature, notably for Baden Württemberg: Maier (1987); Herrigel (1989); Sabel, Kern, and Herrigel (1989); Sabel, Herrigel, Deeg, and Kazis (1989); and for Emilia-Romagna: Brusco (1982, 1990, 1992); Leonardi and Nanetti (1990); Bianchi and Gualtieri (1990); Bianchi and Giordani (1993).

BADEN WÜRTTEMBERG: THE CHALLENGE OF LEAN PRODUCTION

Baden Württemberg is one of Europe's most prosperous and economically powerful regions. In the European Commission's regional rankings, the *Land* scored 120 against the EC index average for GDP per capita of 100 over the 1986–8 period. The Mittlererneckar conurbation, centred upon Stuttgart, where the main industrial complex is located, had a GDP index of 134, ninth highest in the EC over the same period (Commission of the EC 1991).

It is one of Germany's core manufacturing regions, specializing in automotive engineering (employing 237,000), electronic engineering (266,000 employees), and machine-building (including machine-tools) (281,000 employees). Major firms include Daimler-Benz, Porsche, Audi, and Robert Bosch in automotives; SEL-Alcatel, Sony, IBM, and Hewlett-Packard in electronics and leading machinery; and machine-tools firms such as Heidelberg, Trumpf, and Traub.

Larger firms are found in automotive and electronic engineering, with some single plants, such as Mercedes-Benz at Sindelfingen near Stuttgart, with 46,000 employees, being very large. Bosch employ some 75,000 in the Stuttgart area while in the same location, SEL-Alcatel employ some 30,000. These large firms are surrounded by and interspersed with a multitude of small and medium enterprises (SMEs), many of which are their suppliers. The machinery industry is, by contrast, principally composed of SMEs. At the larger end, Trumpf, for example, the world's leading producer of laser-cutting machine-tools, employs only 3,000, a figure reached very rapidly only during the late 1980s. Most machine-tool firms are much smaller, employing typically 100–200 persons. The machine-tool industry is the classic domain of the *Mittelstand* of SMEs. The automotive and electronic engineering sectors display more of a firm-size mix from the very large to the very small.

Until the 1990s Baden Württemberg was seen as a model of the networked economy specializing in diversified quality production (Streeck 1989). The tradition of the 'proud engineer' disposed by training and culture to *pfifferinge* or innovating by tinkering, is bolstered by the hierarchical power of the highly skilled *Meister* who supervise the production process and demand virtually complete autonomy over that sphere. This is true of both large and smaller *Mittelstand* firms, but a particular source of engineering excellence in the latter, and a key factor in the strength and longevity of the *Mittelstand* in the industrial system.

Vertical subcontracting networks function at their strongest between the engineers of the large original equipment manufacturers (OEMs) and those of the *Mittelstand* who, because of their traditional autonomy in both kinds of firms will typically engage in negotiations concerning technological improvements, driven solely by questions of engineering excellence rather than cost. Until recently, firms such as Mercedes-Benz and Porsche have tolerated such luxuries because engineering excellence has traditionally been their key selling point. However, such practices have recently come under the severest questioning.

Lateral subcontracting networks between *Mittelstand* firms are also strong, but constrained in ways that have not always been picked up by some observers (Herrigel 1989; Sabel, Herrigel, Deeg, and Kazis 1989). The point here is that firms in, for example, the machine-tools industry can be surprisingly vertically integrated despite their small size; only subcontract to firms outside their own direct market-niche, i.e. where there is complementarity rather than competition; and are highly protective, as they must be, of proprietorial knowledge. Thus, the collaborative dimension, stressed by Herrigel *et al.*, is less pronounced, especially in production and particularly in machine-tools, than it might appear. This is not to say that such firms are not active supporters and beneficiaries of industry-wide representation through the industry associations and chambers of commerce and industry. As we shall see, these are very active and robust organizations which advise and convey industry concerns within the industrial system and to government on a regular basis.

One of the key issues placing both vertical and lateral networks under duress in the 1990s has been the pressure for firms to adopt 'lean production' (Womack, Jones, and Roos 1990). It is no exaggeration to say that this production model has sent shock-waves of seismic proportions through German manufacturing industry, nowhere more so than in Baden Württemberg. Between 1990 and 1992 there were at least fifty conferences held in Germany on the results of the MIT International Motor Vehicle Programme, stimulated by the findings of the book by Womack *et al.* Key firms such as Daimler-Benz, Volkswagen, and Robert Bosch purchased one thousand copies of the book each as a manual for their top management. The European author, Jones, has made over thirty presentations to the main boards of Daimler-Benz, Volkswagen, Porsche, Continental, IBM (Munich), business associations, and even individual *Mittelstand* firms.

Why this extraordinary interest? First, the study anatomized the Japanese, particularly the Toyota, production system, showing the sources of their automotive industry excellence. Second, Toyota's new luxury car, the Lexus, had been launched as a Mercedes-quality model at half the Mercedes price, a matter of especial concern to the German automotive industry. And third, the book provided a coherent language and focus for numerous manufacturing innovations that German manufacturers had been seeking to integrate, ranging from introducing teamwork to the principles of Computer Integrated Manufacturing (CIM). It could be said that this was a paradigm case of the medium being the message.

In brief, what the 'lean' message was pointing to was the need for 'systems integration' of many of the discrete managerial and production innovations perceived as being Japanese in origin. The key point here is that in isolation they seldom lead to major quality or productivity improvements because they are 'only fully functionable when all elements are in place and working together' (Jones 1990). This is the real challenge of lean production, namely to integrate all of the elements outlined in Table 5.1. Above all, achieving this is a human-resources matter not a technical one, but the ultimate objective is simultaneously to raise quality and reduce costs (on this see Imai 1986; Roth 1992; Sengenberger 1992; Cooke 1992).

The Economic Context

As two of Baden Württemberg's key industries, automotive and machine-tools were assailed by the principles of lean production in 1990/1, so the immediate euphoria of German reunification in 1989 was beginning to

TABLE 5.1. *Key principles of lean production*

Human-oriented principles	Technical principles
Teamwork	Zero defects
Kaizen (continuous improvement)	Zero buffer time (JIT)
Customer orientation	Efficient R&D
Enterprise culture	Supplier integration

Source: after Roth (1992).

TABLE 5.2. *Key economic indicators for Germany, 1991–1992*

Indicator	1991	1992
Real GNP growth (%)	3.2	1.1
Consumer price increase (%)	3.5	3.6
Industrial wage rate increase (%)	6.8	5.4
Industrial production increase (%)	3.0	0.5
Unemployment (%): West Germany	6.3	6.8
East Germany	10.5	14.1
FT-A Index (% change over year)	+6.4	−11.5

Source: *Financial Times*, 26 Oct. 1992.

wear off. By 1992 the German economy had entered recession, as Table 5.2 makes clear. The reasons for this turnaround in German economic fortunes were fourfold:

- Costs: FRG is a high wage, tax, rent, and environmental cost economy. Competitiveness has been undermined to the extent that, strategically, many engineering companies are placing production off-shore while attempting to maintain higher value-added activities such as R&D at home.
- Reunification: the cost of restructuring the former East Germany is presently running at £60 billion per year, and is likely to remain so for the next decade. The inflationary impact of this will ensure that German interest rates will be high for the foreseeable future, hitting both investment and consumption.
- Exports: German exports are expensive, even more so given the devaluations elsewhere in the European Community. Italy and the United Kingdom, for example, are Germany's second and fourth largest trading partners. Exports there and elsewhere may be expected to decline in the absence of some adjustment in German firms' purchasing and location policies.

- Global Competition: increasing competition from Japan and elsewhere in the core industries of automotive engineering and machinery/machine-tools means German firms are in general aiming to increase the value of subcontracted parts from 40% to 70% by 1995. More and more of this increased sourcing will be outside Germany.

The German Automotive Industry

As an indicator of the pressure to respond to competition by moving production into the supplier base rather than producing in-house, Fig. 5.1 and Table 5.3 are instructive. Fig. 5.1 graphs the rise and decline of in-house production by German automotive firms over the 1970–87 period. The graph reveals a sharp increase in in-house production up to around 1980 and thereafter a sharp decline, accelerating rapidly from 1985 onwards. This shows that, although lean production may be a new concept, this element of cost-reduction is one which German automotive firms were increasingly pursuing throughout the 1980s, but with extra determination towards the end of the decade.

Table 5.3 provides information on change in in-house production for each of the key automotive assembly firms in Germany. Of particular interest to the case of Baden Württemberg is the performance of Daimler, a key subsidiary of which is Mercedes-Benz, Baden Württemberg's leading car company. Mercedes was not only an order of magnitude more than the

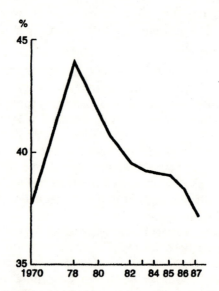

Fig. 5.1. *Production depth in the German automobile industry*
Source: IG Metall, Frankfurt

TABLE 5.3. *In-house production by value in German auto firms, 1970–1987* (%)

	1970	1978	1980	1982	1983	1984	1985	1986	1987
Daimler	41.7	47.4	45.2	43.4	43.7	43.9	44.4	46.9	46.3
BMW	39.5	45.6	42.4	40.2	40.0	40.9	39.7	36.6	35.7
Ford	36.2	42.2	35.1	35.4	34.1	30.5	29.9	31.8	32.9
Opel	36.7	40.7	32.9	31.3	30.5	27.7	28.2	27.6	30.0
Audi	32.0	33.4	33.7	30.1	32.6	30.4	31.7	25.9	29.2
Volkswagen	37.4	35.3	34.4	30.9	29.5	32.0	30.7	28.6	28.6
MEDIAN	38.0	41.1	38.4	36.3	35.7	35.5	35.3	35.0	35.1

Source: IG Metall, Frankfurt.

others in vertical integration, but substantially increased as such over the 1970–87 period. Daimler, and Mercedes especially, is confronted with a serious need to reduce its costs, become much more competitive, and produce a less over-engineered product, something which the firm has belatedly recognized.

The German Machine-Tool Industry

Because machine-tool firms are mostly SMEs they are confronted with the dilemmas of lean production, costs, and competition in a different way. The problem was highlighted in a recent document (Brödner and Schultetus 1992), researched jointly by an industry association and the trade unions, which compared the relative 'leanness' of the German and Japanese machine-tool industries. Taking a representative sample of firms from both countries, the key findings, summarized in Table 5.4, were, from a German viewpoint, fairly sombre. It should be borne in mind that the Japanese and, to a lesser extent, other East Asian producers are seen as the most serious competitors by German producers. More shock-waves passed through the industry when it was announced that Japanese firms such as Citizen had actually bought Baden Württemberg machine-tool companies. The nature of the crisis is essentially that of a huge productivity gap. Japanese firms are larger, massively more productive, and more profitable with proportionately lower labour costs and lower inventory. The problem in machine-tools is that, to compete, those in Germany must concentrate ownership, but the 'proud engineering' tradition means they will not. So, key technologies such as computerized control systems (CNC, etc.), which were being produced in-house, are being out-sourced from large or specialist companies. This leaves the *Mittelstand* firms with the prospect of increasingly becoming 'metal-bashers' rather than innovators.

TABLE 5.4. *Key economic indicators of Japanese and German machine-tool firms*

	Japan 1990 selected firms			German average	Germany 1990 selected firms			
	A	B	C	1989	A	B	C	D
Turnover/employee (DM000)	650	795	725	179	239	283	311	199
Value added/employee (DM000)	336	517	249	95	119	132	149	113
In-house production (%)	52	65	34	51	50	47	48	57
After-tax profits (%)	14	6	8	1.3	-2	2	5	2
Labour costs/turnover (%)	10	12	10	34	31	23	24	38
Stocks/turnover (%)	15	17	22	37	22	20	15	35

Source: Brödner and Schultetus (1992: 39).

The Crisis in Baden Württemberg

The pressures being faced in the automotive and machine-tool industries in Germany are experienced with particular intensity in Baden Württemberg because they are of such relative importance there. Machine-tools employs some 46,000 workers in 391 firms while the automotive industry, with its 237,000 employees, has over 800 supplier firms alone, accounting for 6% of the Baden Württemberg manufacturing economy by value.

Of special interest is clearly the regional response to a looming crisis which has already seen Mercedes-Benz shed 20,000 workers, Porsche 4,500, and Bosch just under 1,000 in 1992 alone. It is widely commented that Germany's car industry must lose 200,000 jobs in the immediate future, according to the German Auto Manufacturers Association (VDA) and some, such as the former head of GM-Opel, think at least twice as many would be nearer the mark.

Institutional Responses

Baden Württemberg has a dense network of intermediary institutions that functions as an industry-support architecture which mobilizes to seek out ways of assisting business. Sometimes, the fine line between legitimate industry–institution links and illegality is transgressed in this process as the demise of Baden Württemberg's celebrated former Minister President Lothar Späth over too-ready acceptance of favours from, for example, the Bosch organization testifies. In the innovation field alone Baden Württemberg boasts:

- 11 Max Planck Institutes for fundamental research
- 13 Fraunhofer Institutes for applied research
- 20 Industrial Contract Research Institutes
- 120 Steinbeis Foundation technology transfer centres for SMEs
- 9 Universities
- 39 Polytechnics

Much of the research activity of these institutions is industry-orientated and funded.

At the level of the *Land* government, the Ministry of Economic Affairs and Technology alone is responsible for:

- Haus der Wirtschaft—Trade and Industry Promotion
- Landesgewerbeamt—Promotion of SME Co-operation
- GWZ—Agency for International Economic Co-operation
- Steinbeis Foundation—Assessment of Venture Funding for SMEs
- Landeskreditbank—State Credit Bank.

Beyond this citadel of largely public institutional support are:

- 13 Chambers of Industry and Commerce
- Employers Associations
- Trade Unions
- Industry Associations.

All of these, in their various forms, have either *Land*, *Kreise* (District), or local level institutional organization and representation. An example of the speed with which information and action can be mobilized through such a network system, the recent case of the *Mittelstand* innovation deficit is illustrative.

One side-effect of lean production is that it creates a new demand for R&D capacity from upper-tier supplier firms from whom more out-sourcing is being required. *Mittelstand* firms, not used to and fearful of the costs of conducting formal R&D, conveyed these fears to the VDMA and VDA (machinery and automotive industry associations at *Land* level). The VDMA took this up with the Ministry of Economics who commissioned Arthur D. Little Inc. of Boston, USA to research the question. The Little report said R&D was needed by *Mittelstand* firms but confirmed their inability to finance it. The Minister of Economics proposed to the *Land* cabinet that SMEs be encouraged, with an incentives package, to collaborate on R&D. After discussion with *Mittelstand* firm representatives who expressed fears about loss of proprietary knowledge if they collabor-ated with competitors, the Ministry proposed that a Fraunhofer Institute should act as the 'honest broker' protecting the proprietary knowledge of the SMEs who were to collaborate. This solution was accepted and the programme implemented: the process, from first identification of the problem to implementation of the solution, took six months.

LARGE-FIRM RESPONSES

Perhaps the best way of indicating how large firms are responding to the stresses they are facing is to summarize the strategies of three leading automotive firms: Mercedes-Benz, Audi, and Porsche. In the case of Mercedes, the first priority has been to get a hold on escalating costs. This has meant infringing the practices of Mercedes engineers making side-deals with supplier-firm engineers on engineering rather than cost grounds. There is a widespread sense that the last new model, the S-class Mercedes, was massively over-engineered and ridiculously expensive as a con-sequence. Moves are being made towards 'simultaneous engineering' on a project basis, where combined teams of R&D, design, and production engineers work with marketing specialists and technical staff of supplier firms to produce a manufacturable product the consumer actually wants. The recent announcement of a move down-market and diversification into off-road and 'people-mover' vehicles bears witness to the new philosophy—entirely in line with 'lean' principles—of listening to the market (Done 1993). Finally, Mercedes now emphasize teamwork and increasingly global modular assembly—also lean principles—in order to raise quality (see, for detail, Morgan, Cooke, and Price 1992).

Audi are ahead of Mercedes in terms of what they call 'synergistic development' of project-based simultaneous engineering teams on the new 80 and 100 ranges. Audi has the aim of reducing in-house production from the present 45% to below 30%, reducing direct suppliers from 1,050 in 1991 to 650 in 1994. Thus, Audi will elaborate a supply-chain and pass more responsibility for quality to fewer first-tier system or modular suppliers who will deal with A-part suppliers, and so on down to component suppliers who deal with parts suppliers. First-tier suppliers are increasingly responsible for R&D and other innovation activities. Suppliers are controlled through 'open-book' accounting whereby the OEM can instantly conduct a value-analysis of total production costs of its suppliers. In general, this shift in emphasis favours the Purchasing over the Engineering department, a matter giving rise to interesting power struggles and alliances (on this also, see Sabel, Kern, and Herrigel 1989).

Porsche has always out-sourced as much as 75% of its value, not least because it is a relatively small, luxury producer without the internal resources to become vertically integrated. The Porsche problem is that the Japanese makers of sports cars can produce high-quality vehicles at half the price or less of a Porsche with new models capable of being introduced by a single firm every three years as against Porsche's ten. Like Mercedes, Porsche has adopted Japanese cost-targeting methods to control its engineers and to begin seeking suppliers on a far more global basis than its normal Zuffenhausen backyard. The weakness of Porsche's internal revenue generation capacity suggests that it may seek full acquisition from its Volkswagen–Audi shareholding interest if it cannot ride out the present disastrous, particularly for the likes of Porsche, recession.

The Mittelstand *Response*

The Baden Württemberg *Mittelstand* is transfixed rabbit-like in the glare of the oncoming Japanese competitive threat. On the one hand, there are world number one firms, specialized in particular technologies, such as laser-cutting, who are relatively unconcerned about the niceties of lean production. On the other there is an underclass of jobbing parts, components, and tooling firms, perhaps increasing; flexible in the extreme, unencumbered by the requirements of innovation, living what may soon be a hand-to-mouth existence dependent upon how many orders their *Industrievertretungen* or agents can find for them. In the car industry, Walter Reister, the Baden Württemberg leader of IG Metall, the

Metalworkers Union, predicts that the present 3,500 German suppliers will reduce to 1,000 by the end of the present restructuring round (see for detail, Cooke, Morgan, and Price 1993). In the machine-tool industry, a broadly similar future can be expected unless firms actually become more favourably disposed to collaborative competition, the formation of groups, and ultimately fusion as an alternative to acquisition.

One possible pointer to the future for *Mittelstand* machine-tool firms is that proferred by the example, noted earlier, recommended by the Little report. This is now being extended to the machine-tool industry, having been first implemented in automotive components, under the new coalition CDU–SPD (Christian Democrat–Social Democrat) *Land* government that took power from the CDU in 1992. If this line is pursued, there is no better exemplar to learn from than the AKZ consortium of Baden Württemberg engineering firms, about thirty in number, who have successfully collaborated on research, design, marketing, and common-purchasing for the past fifteen years. However, to return to a key theme of this discussion of the industrial system in Baden Württemberg, the nature of the collaborative effort, the key reason why the AKZ collaboration has been successful and long-lasting is that none of the firms are direct competitors and many of them possess 'complementary assets' (Teece 1986).

To summarize the analysis presented for the Baden Württemberg case, the following three key points bear repetition. First, Baden Württemberg is a robust, prosperous region with a well-embedded industrial support system which takes the form of a network architecture that is alert to crisis and responsive to change. This system is presently under duress but by no means in distress. Second, the principal reason why the Baden Württemberg system is under duress is that it has, like the rest of former West Germany, been assailed by the combination of agglomeration diseconomies, the costs of German reunification, *Bundesbank* anti-inflation policy which has helped induce recession, and the pressures of lean production *à la* Japan. Finally, out of this concatenation of economic misfortunes may be emerging a new 'Modell Baden Württemberg' in which the possible advantages of collective entrepreneurship are really beginning to be explored.

THE EMILIAN MODEL: STRESS OR CRISIS?

Few regions have exerted as much influence on local economic development thinking in the past decade as Emilia-Romagna in central Italy, which is celebrated on both sides of the Atlantic as an example of a dynamic and flexible SME-based regional economy. It is not difficult to see why Emilia-Romagna has attracted such international attention: it ranks eighth in the league of European regions in terms of per capita income, it had one of the fastest growth rates in Italy in the 1980s, and in 1991 its unemployment rate was just 3.8% against 10.8% for Italy as a whole.

This is no mean achievement for a region which is based on so-called 'mature' industries, e.g. mechanical engineering, textiles, agriculture, food, furniture, and ceramics. With the exception of machine tools none of these industries would fall into the so-called 'high technology' bracket. However, as we can see from Table 5.5, the region's main industries recorded impressive growth rates during the 1980s.

Apart from this 'mature' industrial base another distinctive feature of the regional economy is the pronounced bias towards small firms. As we can see from Table 5.6 nearly 75% of all employment in 1988 was in firms with less than three employees. This small-firm bias has been both a strength and a weakness at various points in the evolution of the regional economy. To illustrate this point let us look at the various phases through which the Emilian model has evolved. The traditional Emilian model had a number of features, including:

TABLE 5.5. *Index of production in selected industries in Emilia-Romagna*
(1979 = 100)

	1980	1984	1987	1988
Foodstuffs	95.4	126.6	219.3	282.9
Textiles	97.1	107.8	171.6	195.2
Metalworking	100.8	92.7	113.2	133.6
Furniture	99.3	89.4	91.3	106.0
Chemicals	99.0	91.4	133.1	137.3
Ceramics	100.1	106.2	110.7	127.1

Source: Regional Chamber of Commerce.

TABLE 5.6. *Firms and employees in Emilia-Romagna, 1988*

Employees	Firms	%
0–2	230,023	74
3–9	61,788	20
10–49	15,508	5
50–99	1,222	0.4
100–499	847	0.3
› 500	71	0.2
TOTAL	309,459	100

Source: CERVED–SAST.

- a dense network of SMEs in subcontracting relationships: some firms specialized in part of the production process, while others were responsible for the finished product. This system was said to be governed by a highly distinctive social regime of rules and regulation which set a high premium on trust, with the result that opportunism was frowned upon;
- local specialization in industrial districts: which meant that textile production was concentrated in Carpi, ceramic tiles in Sassuolo, luxury cars in Modena, etc.;
- a robust system of institutional support: the most novel feature being the decentralized business service centre system which is engaged in the provision of 'real services' to local firms (Brusco 1982; Piore and Sabel 1984; Best 1990).

The Evolution of the Districts

Following Bianchi and Gualtieri (1990) it is possible to identify three distinct phases in the evolution of Emilia's industrial districts. Phase one begins in the 1960s and runs to the end of the 1970s. Emilia-Romagna, like many northern Italian regions, had a large population of small firms supplying basic products for a local market. A case in point would be the agricultural machinery and repair firms, which developed alongside the large agricultural and foodstuff industries. Many of these small firms were artisan firms which were nurtured by very supportive artisans' associations such as CGA (General Confederation of Artisans) or CNA (National Confederation of Artisans). In the 1960s large firms throughout Italy were seeking buffer-suppliers due to increasing variations in demand. Artisan firms in the machinery industry were targeted by these large firms as a means of dealing with quantitative instability.

In textiles, however, the trajectory was different again. Carpi, the main textile district, had begun as a centre of straw-hat production. When this industry collapsed in the 1940s agricultural homeworking—the main source of production—dried up. However, artisans used their contacts with large firms to begin selling (not making) cheap clothing. Later, some began producing T-shirts using the homeworking tradition in the area. As agricultural work declined, more labour was taken on in small firms which began to expand into factory production. Whether directly or indirectly large firms influenced the course followed by SMEs in this first phase.

In the second phase, from the late 1970s through to the 1980s, those SMEs which had established themselves as subcontractors or independent producers increased both the quantity and the quality of their output. This phase coincided with a major crisis of the large-firm sector in the early 1980s, a crisis which was partly induced by the fact that the market for standardized goods was becoming ever more competitive as a result of the focused export drives on the part of newly developing countries. While many large firms simply vacated price-sensitive standardized product markets, like clothing for example, Emilian SMEs were able to respond faster and more flexibly to the changing market conditions. In a sense then, the success of the Emilian SMEs was facilitated by the crisis of the large-firm sector in Europe.

The third phase in the evolution of the Emilian SME sector has been under way since the beginning of the 1990s. Without doubt it is the most difficult period faced by the firms of the industrial districts, provoked in part by more aggressive and more flexible strategies on the part of the large firms on the one hand and, on the other, by increased competition from low-wage countries. Faced with heightened global competition large firms have responded in ways comparable to SMEs a decade earlier, i.e. they have developed more flexibility in their management of the product cycle. Instead of producing for the mass market they have focused upon more rapid turnaround in more diversified quality-conscious markets.

An innovation of particular importance has been the move by large firms in the machinery, garments, and food-processing industries to control distribution networks. This is consistent with one of the effects of economic integration and the globalization of production, namely the need to gain access to the markets of competitor firms. Frequently, this is achieved through the formation of complementary strategic alliances or joint-ventures with other large firms. SMEs find this form of competition very difficult to combat because, in many cases, they are price-takers and product-followers, i.e. they react to the innovations of others. When price and product leaders leave space, this works. But if large firms are seeking to introduce oligopolistic rule—by exerting stronger control of distribution chains, for example—this makes life far more difficult for SMEs. This is the phase in which the Emilian SME sector now finds itself.

In what follows we propose to look at two interrelated aspects of this third phase. First we shall examine some current economic restructuring trends, using the clothing and engineering sectors to illustrate the problems and the challenges for the region in the 1990s. Secondly, we shall examine the political response to this problematical new phase and here we shall argue that the region's prospects depend in no small way on a radical reform of the national institutional fabric, e.g. the division of labour between national and regional government, the relations between industry, banks, and higher educational institutions.

Economic Restructuring: The Challenges Ahead

The key economic challenges facing the regional economy can be summarized very quickly:

• like many other regions in northern and central Italy Emilia-Romagna has seen its comparative price advantage eroded by increased wages, social security, and general production costs;

● many of the smaller, family-owned firms are ill-equipped to finance the twin demands of new technology (which is essential for higher-quality production and reduced time to market strategies) and global markets (which require more sophisticated marketing and after-sales servicing);

● the region's local subcontracting networks, a source of strength in the past and a core feature of the Emilian model, are beginning to unravel as production is shifted to lower-cost regions and countries, and this poses severe problems for the artisan sector which has specialized in this basic production work in the past;

● new corporate hierarchies appear to be emerging in the region's industrial districts as a result of a growing concentration of capital through especially mergers and takeovers;

● a key problem for the small-firm sector in particular is the relative absence of flexible financial instruments to finance innovation and growth because the risk capital market is underdeveloped both nationally and regionally;

● small firms in all sectors will have to pay far more attention to R&D and product innovation in the 1990s because in the 1980s they tended to focus on cost-cutting and searching for new markets (Cooke and Morgan 1991, 1992).

Although these problems exist to varying degrees throughout the regional economy they appear to be most acute in the clothing industry, a mature sector which is highly populated with very small firms which compete on the basis of price. According to CITER, the Carpi-based business service centre, some 11,136 clothing and knitwear firms operate within the region, 2,600 of which are located in Carpi. Of the Carpi total some 1,850 operate solely as subcontractors, each specializing in a small part of the production process, e.g. knitting fabric, overlocking, button-holing. The remaining 750, which tend to be the larger firms, are in direct contact with the retail market and these firms are typically involved in a much wider array of operations (e.g. design, marketing, grading, pattern-making, packaging, and dispatch), though production tends to be farmed out to subcontractors.

In recent years a process of concentration through merger and acquisition has been evident in the Carpi district, with the result that new corporate leaders are beginning to emerge. In some cases new vertical networks have begun to replace the traditional horizontal networks of inter-firm collaboration in the district. Although this in itself need not spell disaster for the smaller firms the latter are troubled by the new production strategies of the larger firms. For example, some of the larger firms in the district have begun to shift production to low-wage areas within Italy (like Puglia and Calabria) and to low-wage countries in Europe (like Greece and, before the war, in Yugoslavia). The scale of this process of 'offshore production' would be that much greater were it not for the fact that subcontracting beyond the region carries its own problems, like quality assurance and reliability of delivery.

Although it is difficult to assess how far this process has gone there can be no doubt that many of Emilia's market-facing firms in this industry are beginning to question the wisdom of using the region as their main production base. Some local experts (like Patrizio Bianchi of Nomisma) believe that a new division of labour is emerging, in the sense that production will continue to move out of the region, leaving higher-added-value activities like design within the region. If this is a solution for the larger firms, there remains the problem of what is to be done about the smaller firms whose subcontracting market looks set to decline.

Understandably, the regional authorities are deeply concerned about this issue and, as we shall see later, they are trying to animate new forms of corporate collaboration as an alternative to a destructive market-induced form of restructuring. Indeed, the firms themselves have begun to pioneer new forms of corporate collaboration, a trend illustrated by the formation of new groups. These groups tend to be organized across the lines of specific market segments, e.g. men's fashion, female fashion, or sports wear. The key to the reorganization of production within the group lies in the fact that each member assumes responsibility for a specific corporate task (be it marketing, administration, financial management), with the result that other members are able to concentrate on a narrower range of core functions. In addition to these benefits, both suppliers and customers benefit from the bulk-ordering which is beyond the realm of each individual firm. In short, group-formation is both a market-widening and a risk-spreading activity. The largest of these groups has over fifty member companies under the name of Bassoli Pronta Moda. Other groups are MIT and Americanino (which is headquartered in Vicenza outside the region, but consists of mainly Carpi firms).

If group-formation signals a disposition to collaborate to achieve mutually beneficial ends, this collaborative ethos should not be exaggerated. Indeed, with the diffusion of new technology in the Carpi knitwear district it is apparent that early adopters see this technology as a means of creating 'network leadership' roles for themselves within the chain of production. The diffusion of computer-aided design (CAD) technology is a case in point. Through CITER local firms are being offered a locally designed CAD system called CITERA, a facility which leads to faster production of new styles and samples and significant cost-savings. In addition, CAD enhances a firm's responsiveness to market trends and retailers' demands and CITERA users see CAD as a vital part of their strategy to move into higher-value markets.

Some of the early users of the CITERA system have adamantly refused to share the benefits of CAD with their subcontractors, despite the fact that this would enhance both versatility and quality. Although CITERA is capable of producing disks which can be read by programmable knitting-machines of the type used by many subcontractors, these CITERA owners refused to allow this link-up on two counts: first, they feared that suppliers might pirate their designs and, second, they felt that proprietorial use of the system conferred 'network leadership' benefits which might be diluted it if was shared with these suppliers (Cooke and Morgan 1992). This proprietorial attitude to new technology could well undermine one of the key features of the traditional Emilian model, namely a shared technical discourse which was central to the integration and coordination of a decentralized system of production. If this is the case then the task of promoting more innovative forms of corporate collaboration in the 1990s might prove to be more difficult than is often imagined. Against this, however, it is difficult to imagine that informal knowledge of the functioning of technologies is not easily acquired and diffused in such an exchange-rich setting.

If we turn from knitwear to machine tools it is clear that new forms of corporate collaboration will have to be developed here too. The Italian machine-tool industry went from strength to strength in the 1980s, so much so that by the end of the decade Italy had become the second largest EC exporter and the fourth largest worldwide. The success of the Italian industry owes much to its capacity for flexibility and versatility. Italian firms have gained a worldwide reputation for providing tailor-made solutions in the sense that machines are designed and built to satisfy the nuanced requirements of particular customers. According to UCIMU (the Italian machine-tool trade association) the flexibility of the industry is partly a function of its scale: over 80% of the industry's 450 firms have payrolls of less than one hundred workers.

Emilia-Romagna is one of the premier centres of the Italian machine-tool industry, along with Lombardy, the Veneto, and Piedmont. Like their counterparts in these other regions, the leading figures in the Emilian machine-tool sector fear that the current organization of the industry is too fragmented to meet the challenges of the 1990s. There is a growing concern that Emilia's machine-tool firms may be too small to finance new rounds of R&D expenditure, too small to support global market strategies, and too small to produce at the price levels of their Japanese competitors, many of whom have the advantage of long production runs. For all these reasons leading figures in the region—like Cesare Manfredi, who runs a milling machine firm in Reggio Emilia and who is also chairman of UCIMU—are trying to foster greater collaboration, either through mergers or joint-ventures on technical and commercial collaboration. Along with its sister organizations throughout the EC, UCIMU is preaching the message that better inter-firm collaboration is necessary throughout the EC if SMEs are to prosper in the machine-tool market of the 1990s (CECIMO 1992).

The stresses and strains which we have identified in these two industries would seem to be evident throughout the regional economy. Indeed, one well-informed commentator argues that:

The evidence points toward the emergence of deep structural instabilities in certain of the 'system defining' institutional arrangements that are thought to be central for the social reproduction of the districts. The stability of the districts is being compromised by a combination of their growing interactions with the outside world and the evolution of highly uneven power relations within the districts themselves. Even where economic growth continues, the collaborative character of the districts is changing. I maintain that these changes are in a direction *away* from a model of small firm-led regional economic development (Harrison 1993).

Like Harrison, we are not saying that Emilia-Romagna is about to be engulfed by a major crisis; what we are saying, however, is that new corporate and political strategies are urgently required if the region is to maintain a leading position in the Europe of the 1990s.

The Institutions of Collective Entrepreneurship

The success of Emilia-Romagna owes much to its capacity for collective entrepreneurship, i.e. the disposition to collaborate to achieve mutually beneficial ends. This is evident in the corporate realm, where it manifests itself in close inter-firm relations, strong business associations, etc., and in the political realm, where a high premium has been placed on creating a robust and decentralized system of institutional support (Best 1990). The main aim of this institutional support system is to ensure that commercial and technical knowledge is diffused as widely as possible throughout the region, a critically important task in an SME-based economy. Since it was created in the 1970s the regional government in Emilia-Romagna has endeavoured to build a support system tailored to the needs of its SMEs. In view of the limited powers and scarce resources available to the Italian regions it is a tribute to the Emilian authorities that they have achieved so much with so little.

At the heart of the region's institutional apparatus lies ERVET (Ente Regionale per la Valorizzazione Economica del Territorio), the regional development agency established in 1974. The shareholding structure of this agency reflects an ethos of collaborative engagement in the sense that ERVET has tried to integrate the potential of public and private sectors, of credit and financial institutions, entrepreneurial associations, and chambers of commerce, etc. Although ERVET is itself the dominant shareholder it sets a high premium on involving as wide a social constituency as possible. The reasons for this are twofold: first, budgetary constraints mean that all possible sources of finance must be tapped and, second, the involvement of private-sector interests means that the latter have a direct stake in the support system, which in turn enhances its credibility in the eyes of local firms, the main targets of the support system.

ERVET provides its services either directly or through its wider support network, a network which consists of two parts: the service centre system and the enlarged system (see Fig. 5.2). Some of the service centres are sectoral in focus (like CITER, the Carpi-based business support centre for the clothing and textile industry), while others have a multi-sectoral focus (like the Bologna-based ASTER which is a technology transfer centre aimed at the whole region). Although this regional support system has been extolled as a model of collective service provision by many foreign observers, the fact is that this system was rarely evaluated in terms of the effect it had on the region's SMEs (Bellini *et al.* 1990; Cooke and Morgan 1991).

In the wake of the 1990 regional elections, and the advent of a coalition government, a major evaluation exercise was launched to examine the actual effectiveness of the system. Although the review was mainly stimulated by a crisis of public finance (an acute problem for the nation as a whole), a review was also considered necessary to enable the region to meet the new challenges of the 1990s—not least the emergence of new corporate hierarchies as a result of takeovers, group-formation, and internationalization. One of the key dilemmas here is the 'increasing asymmetry between the competence of the regional government authority and the physical extension of financial, technological and productive spheres to which the companies in Emilia-Romagna are ever more closely linked' (Bellini *et al.* 1990: 136). In other words, the regional support system, designed to sustain locally owned and locally based firms with a strong commitment to their locales, is having to adjust to a new economic environment, one which is less conducive to regional animation.

Although the results of the review have yet to be fully implemented, it is already clear that a major process of institutional innovation is under way. In strategic terms the ERVET system is being reorganized so that it is in a better position to promote three principal activities, namely innovation, internationalization, and finance, each of which is deemed to be crucial to the future health of the regional economy.

On the innovation front the regional authorities are proposing that the ERVET system becomes much more integrated with the new regional technological poles (at Bologna, Parma, and Piacenza) so that the benefits of applied research projects can be diffused more widely throughout the region. In addition ERVET has been enjoined to develop synergies with the regional university system because these links are relatively under-developed.

As regards internationalization the regional authorities feel that there is no alternative to the growing integration of the regional economy into the wider structures of the European Community. Indeed, this link with the

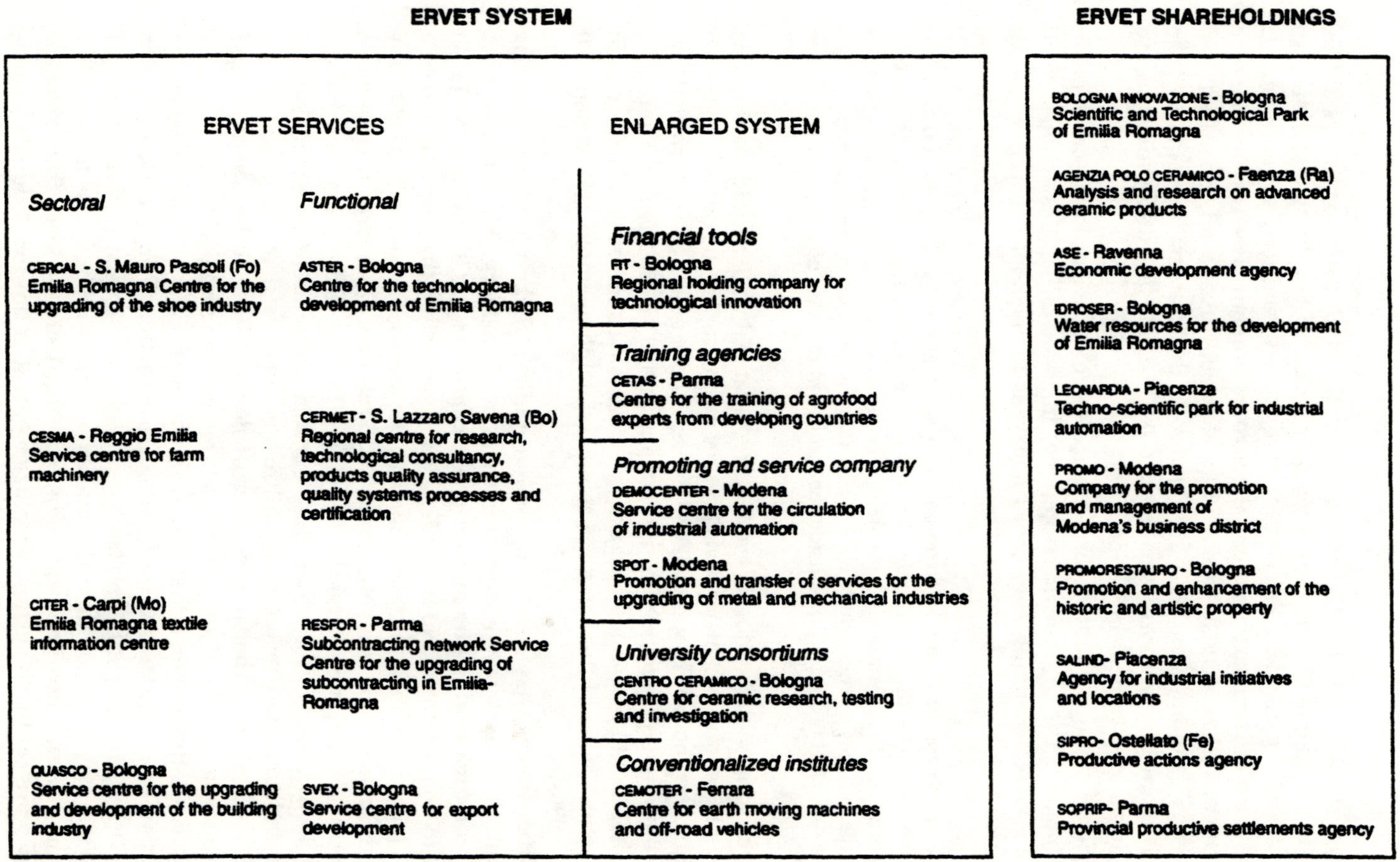

ERVET SYSTEM

ERVET SERVICES

Sectoral

CERCAL - S. Mauro Pascoli (Fo)
Emilia Romagna Centre for the upgrading of the shoe industry

CESMA - Reggio Emilia
Service centre for farm machinery

CITER - Carpi (Mo)
Emilia Romagna textile information centre

QUASCO - Bologna
Service centre for the upgrading and development of the building industry

Functional

ASTER - Bologna
Centre for the technological development of Emilia Romagna

CERMET - S. Lazzaro Savena (Bo)
Regional centre for research, technological consultancy, products quality assurance, quality systems processes and certification

RESFOR - Parma
Subcontracting network Service Centre for the upgrading of subcontracting in Emilia-Romagna

SVEX - Bologna
Service centre for export development

ENLARGED SYSTEM

Financial tools

FIT - Bologna
Regional holding company for technological innovation

Training agencies

CETAS - Parma
Centre for the training of agrofood experts from developing countries

Promoting and service company

DEMOCENTER - Modena
Service centre for the circulation of industrial automation

SPOT - Modena
Promotion and transfer of services for the upgrading of metal and mechanical industries

University consortiums

CENTRO CERAMICO - Bologna
Centre for ceramic research, testing and investigation

Conventionalized institutes

CEMOTER - Ferrara
Centre for earth moving machines and off-road vehicles

ERVET SHAREHOLDINGS

BOLOGNA INNOVAZIONE - Bologna
Scientific and Technological Park of Emilia Romagna

AGENZIA POLO CERAMICO - Faenza (Ra)
Analysis and research on advanced ceramic products

ASE - Ravenna
Economic development agency

DROSER - Bologna
Water resources for the development of Emilia Romagna

LEONARDIA - Piacenza
Techno-scientific park for industrial automation

PROMO - Modena
Company for the promotion and management of Modena's business district

PROMORESTAURO - Bologna
Promotion and enhancement of the historic and artistic property

SAUND- Piacenza
Agency for industrial initiatives and locations

SIPRO - Ostellato (Fe)
Productive actions agency

SOPRIP- Parma
Provincial productive settlements agency

FIG. 5.2. *ERVET system and shareholding, 1990*
Source: ERVET

EC is perceived as a means of preserving the vitality of the region's SMEs. As the Regional Commission argued: 'whether or not the size of SMEs can be maintained, given the current level of competition, depends on the existence of (and participation in) networks of relations which operate on a regional rather than a local level and which, above all, open out into Europe' (Commission 1992: 23). In view of the strategic importance of the EC the regional authorities have given ASTER the task of coordinating the region's involvement in EC networks.

Finally, major initiatives are deemed to be necessary in the crucial field of finance. The Commission noted that financial difficulties amongst small firms were becoming ever more apparent throughout the region and it argued that:

these problems are largely attributable to the historical failings of our financial system as regards the supply of venture capital and of merchant banking services . . . Furthermore, on the demand side, businesses have on the whole been hostile towards opening share capital to actors outside the controlling group (in the case of large businesses) or outside the family (in the case of SMEs) (Commission 1992: 23).

After some deliberation it was decided that ERVET should not become directly involved in the provision of venture capital to SMEs; instead it should be the primary interface between the region's SMEs and national and international sources of investment capital. In short, new sources of capital had to be tapped because the traditional sources within the region were incapable of financing the commercial and technological demands of the 1990s.

The review has also proposed a radical shake-up of the existing institutional system: some centres are to be closed (e.g. SPOT), others are to be merged, while the most effective centres (e.g. CITER and ASTER) are to be reinforced. No less radical is the suggestion that there should be a 'progressive reduction of subsidies' in the ERVET system. That is to say there seems to be growing support for the idea that, over a reasonable period of time, most services should aim to be self-financing, and indeed profit-making. On the other hand the regional authorities admit that too few indigenous firms actually utilize the services that are on offer, the implication being that new sources of demand have to be found, in and beyond the region, if the service centres are to remain viable.

It is in this context that the regional government is considering the possibility of 'deregionalizing, if not internationalizing, certain service activities', a strategy that is clearly only open to the best service centres. If this strategy succeeds then the possibility arises that centres like CITER might find that their interests begin to diverge from the local districts which they were set up to sustain. In other words it is not just Emilia's leading firms that are thinking about rebalancing their activities as between the region and the wider world: parts of the regional institutional apparatus are also considering this option.

Whatever the fate of these reforms it is patently clear that the process of institutional innovation within Emilia-Romagna is not just a regional affair. For example, the regional authorities have set a high premium on forging synergies between local industry on the one hand and the banking and university sectors on the other. Yet from a local economic development standpoint the potential of both these sectors has been vitiated by the fact that they have been unable to engage with local industry to anything like the same extent as in other European regions. The antiquated Banking Act of 1936 prevented local banks from developing equity partnerships with local industry. Similar limits on the higher education sector, which was controlled to a suffocating degree by central government, meant that universities had little opportunity to forge partnerships with their local economies.

In other words dynamic regions like Emilia-Romagna have until now been constrained by antiquated national regulations. Although these regulations are being reformed to enable both banks and universities to engage more forcefully with local industry, it is remarkable that Emilia-Romagna has achieved so much under such a national system. However, as we move into an era of systemic innovation, which requires strong synergies between industry, finance, and the science and technology base, Emilia-Romagna is not as well placed as regions like Baden Württemberg, where these synergies have existed for much longer.

The restructuring of the Italian state system, triggered by the crisis of public finance on the one hand and the malaise of the ruling political coalition on the other, may be a blessing in disguise for the stronger Italian regions. If a more pronounced regionalized state system were to emerge from the current national crisis then regions like Emilia-Romagna would be well placed to benefit from a system which devolved financial and industrial powers to the regional level. This is the wider institutional challenge which has to be met if the regional innovation process within Emilia-Romagna is to stand any chance of long-term success.

CONCLUDING REMARKS: COPING WITH COMPETITION?

The two regions we have discussed are remarkably different in their economic profiles but have some quite significant features in common with respect to: the nature of their present crises, the origins of the particular forces undermining their recent accomplishments, and the forms of political and policy initiative being taken to restore their fortunes.

Baden Württemberg's is primarily an engineering economy, one in which the very highest reaches of excellence have been attained. Of crucial importance to that achievement has been a long-established network architecture of collective support of the industrial base through expert intermediary institutions. This delicately balanced industrial complex has been destabilized by the direct, head-to-head, competitive effects of Japanese lean production methods. Emilia-Romagna is a relatively youthful SME-dominated economy, the prosperity of which has been earned through the judicious application of the principles of a decentralized collective entrepreneurship and the ingenuity of the social base which itself practised as well as gave political legitimacy to that form of enterprise.

Global economic pressures are felt in Emilia-Romagna almost as much from the Third World as from, say, the EC, not least because many of its industries are sophisticated but relatively low value-added. Clothing, shoes, furniture, and even ceramics are particularly prone to outside competition from less-developed parts of the world. Only in Emilia's higher-value food-machinery, earth-moving, and machine-tool industries is this less the case. There, more of a threat comes from the international commercial pressures associated with the Single Market. In Baden Württemberg, the pressures are global in the sense of there being a new production model and enhanced competition, primarily from Japan, in its core engineering industries. But internal strains from reunification and anti-inflation policy are even more debilitating.

Political change has occurred in both regions. From being solely controlled by the Communists for a generation, Emilia-Romagna is now governed by a PDS (the reformed Communist Party)–Socialist coalition. This has meant new attention being paid to established forms of network support institutions in the light of both the need to control public expenditure and the globalizing effects of lean production and the Single Market. In the case of Baden Württemberg, the prospects for encouraging collaboration are being further explored, based, to some extent, on experience arising from the implementation of the Little report's recommendations. In the case of Emilia-Romagna a far-reaching inquiry into the adequacy and effectiveness of the whole range of regional and local business support infrastructure has been embarked upon.

In conclusion, while it is unlikely that the practices of intervention to assist industry—especially small-firm support—will be dismantled completely, it is clear that the next phase of institutional development will focus centrally upon the issue of innovation. In the Baden Württemberg case, it is clear that the cost-demands being placed on *Mittelstand* firms by their customers are such that only judicious public partnership can enable some to satisfy the drive for innovation. Equally, in Emilia-Romagna, the demands exerted by market forces upon SMEs now mean that the existing business support centres are too small. More innovative input demands some concentration of limited resources and greater sophistication in the services offered. In both cases, the exigencies of heightened global competition are creating significant challenges to the ingenuity and capability of the regional networks on which prosperity may increasingly depend.

NOTE

The research reported here arises from a project funded by ESRC Grant R000 23 2962 entitled 'Regional Innovation in Europe'. We would like to acknowledge the work of Adam Price, research assistant to the project, especially for translation of key German documents. Others who have assisted our efforts enormously in Baden Württemberg include Gerd Becher and Stefan Kuhlmann of Fraunhofer-ISI, Karlsruhe; Gunter Meyerhofer of VDMA, Stuttgart; Dieter Klumpp of SEL, Stuttgart; Dietrich Munz, Ministry of Economics, Baden Württemberg. All managers of firms we interviewed are thanked for giving freely of their time. In Emilia-Romagna, our work benefited enormously from, amongst others, Patrizio Bianchi and Romano Prodi of NOMISMA in Bologna; Nicola Bellini, economic advisor to the President of Emilia-Romagna; Anna Flavia Bianchi and Paulo Bonaretti of ASTER; the staff of CITER and all the firm-managers whom we interviewed. None bears any responsibility for what we have written.

REFERENCES

Aydalot, P. (1986), *Milieux innovateurs en Europe* (Paris: GREMI).

Bellini, N., Grazia-Giordani, M., and Pasquini, F. (1990), 'The industrial policy of Emilia-Romagna: The business service centre', in Leonardi and Nanetti (1990).

Best, M. (1990), *The New Competition: Institutions of Industrial Restructuring* (Cambridge: Polity).

Bianchi, P., and Gualtieri, G. (1990), 'Emilia-Romagna and its industrial districts: The evolution of a model', in Leonardi and Nanetti (1990).

—— and Giordani, M. (1993), 'Innovation policy at the local and national levels: The case of Emilia-Romagna', *European Planning Studies*, 1: 25–41.

Brödner, P., and Schultetus, W. (1992), *Erfolgsfaktoren des japanischen Werkzeugmaschinenbaus* (Eschborn: RKW).

Brusco, S. (1982), 'The Emilian model: Productive decentralisation and social integration', *Cambridge Journal of Economics*, 6: 167–84.

—— (1990), 'The idea of the industrial district: Its genesis', in F. Pyke, G. Becattini, and W. Sengenberger (eds.), *Industrial Districts and Inter-firm Cooperation in Italy* (Geneva: International Institute for Labour Studies).

—— (1992), 'Small firms and the provision of real services', in F. Pyke and W. Sengenberger (eds.), *Industrial Districts and Local Economic Regeneration* (Geneva: International Institute for Labour Studies).

CECIMO (1992), *The Situation of the European Machine-Tools Market* (Brussels: CECIMO).

Commission (1992), *Review of the ERVET System* (Bologna: Government of Emilia-Romagna).

Commission of the European Communities (1991), *The Regions in the 1990s* (Brussels: CEC).

Cooke, P. (1992), 'The Experiences of German Engineering Firms in Applying Lean Production Methods', 'Lean Production and Beyond' Report (Geneva: International Institute for Labour Studies).

—— and Morgan, K. (1991), *The Intelligent Region: Industrial and Institutional Innovation in Emilia-Romagna* (Regional Industrial Research Report, 7; University of Wales, Cardiff).

—— —— (1992), *Regional Innovation Centres in Europe* (Report to the Department of Trade and Industry, University of Wales, Cardiff).

—— —— and Price, A. (1993), *The Future of the Mittelstand: Competition Versus Cooperation* (Regional Industrial Research Report, 13; University of Wales, Cardiff).

—— Moulaert, F., Swyngedouw, E., Weinstein, O., and Wells, P. (1992), *Towards Global Localisation* (London: UCL Press).

Done, K. (1993), 'Mercedes-Benz slaughters a sacred cow', *Financial Times* (27 Jan.), 1.

Harrison, B. (1993), *Big Firms, Small Firms: Corporate Power in the Age of 'Flexibility'* (New York: Basic Books).

Herrigel, G. (1989), 'Industrial order and the politics of industrial change: Mechanical engineering', in P. Katzenstein (ed.), *Industry and Politics in West Germany* (Ithaca, NY: Cornell University Press).

Imai, M. (1986), *Kaizen: The Key to Japan's Competitive Success* (New York: McGraw Hill).

Jones, D. (1990), 'The further development of the Toyota production system: The age of lean production', paper to International Operations Management Association Conference, Warwick (June).

Leonardi, R., and Nanetti, R. (1990) (eds.), *The Regions and European Integration: The Case of Emilia-Romagna* (London: Pinter).

Maier, H. (1987), *Das Modell Baden Württemberg* (Berlin: IIM (Research Unit Labour Market & Employment)).

Morgan, K., Cooke, P., and Price, A. (1992), *The Challenge of Lean Production in German Industry* (Regional Industrial Research Report, 12; University of Wales, Cardiff).

Piore, M., and Sabel, C. (1984), *The Second Industrial Divide: Possibilities for Prosperity* (New York: Basic Books).

Roth, S. (1992), *Japanization or Going our own Way? New 'Lean Production' Concepts in the German Automobile Industry* (Frankfurt: IG Metall).

Sabel, C., Kern, H., and Herrigel, G. (1989), 'Collaborative manufacturing: New supplier relations in the automobile industry and the redefinition of the industrial corporation', mimeo, Sloan School of Management, MIT, Cambridge, Mass.

—— Herrigel, G., Deeg, R., and Kazis, R. (1989), 'Regional prosperities compared: Massachusetts and Baden Württemberg in the 1980s', *Economy and Society*, 18: 374–404.

Sengenberger, W. (1992), 'Lean Production: The Way of Working and Producing In the Future', 'Lean Production and Beyond' Report (Geneva: International Institute for Labour Studies).

Stöhr, W. (1990) (ed.), *Global Challenge and Local Response: Initiatives for Economic Regeneration in Contemporary Europe* (London: Mansell).

Streeck, W. (1989), 'Successful adjustment to turbulent markets: The automobile industry', in P. Katzenstein (ed.), *Industry and Politics in West Germany* (Ithaca, NY: Cornell University Press).

Teece, D. (1986), 'Profiting from technological innovation: Implications for integration, collaboration, licensing and public policy', *Research Policy*, 15: 285–305.

Womack, J., Jones, D., and Roos, D. (1990), *The Machine that Changed the World* (London: Macmillan).

6

Flexible Districts, Flexible Regions? The Institutional and Cultural Limits to Districts in an Era of Globalization and Technological Paradigm Shifts

Amy Glasmeier

INTRODUCTION

Since the early 1980s scholars have been intrigued with small-firm industrial production complexes. The resulting literature has emphasized the viability, flexibility, and longevity of this form of industrial organization (Sabel 1989; Scott 1988; Storper and Walker 1990). In large part this literature was intended as an antidote to large-firm dominance in much scholarly work on regional and international development. With greater reflection, scholars are now beginning to question the efficacy of small-firm production complexes (Sayer 1989; Harrison 1993). They are recognizing that their existence is both time- and sector-dependent, and therefore industrial districts may represent only a limited model of future regional development (Storper and Harrison 1991; Glasmeier 1991).

This chapter adds to the growing debate about the efficacy of small-firm clusters by suggesting that there are costs as well as benefits attached to the local agglomeration of an industry. As the following story of the Swiss watch industry notes, localized network systems are particularly vulnerable to external shocks in the form of paradigm shifts. Individual self-interest and limited information flows can inhibit a local complex's ability to transform institutions that otherwise regulate industrial networks. These institutions often have only a limited ability to effect change and are more often a reflection of the complex rather than a window on to the outside world.

The following case study attempts to show the difficulties that confront highly disintegrated production systems in a period of technological change. It is both a story of great success, and ultimately a story of institutional inertia. More than two hundred years ago a precision industry, watchmaking, lodged itself in a rural region of Switzerland to create a technological complex of unusual proportions. Unlike its sparsely populated regional counterparts, the Jura defied conventional wisdom. It was a technological leader among remote regions.

How did this remote mountain region gain world market leadership? The highly vertically disintegrated production structure promoted intense specialization, resulting in the formation of superior regional skills. In the early life of the industry, this was a great advantage. The regional production complex was self-contained, and both watches and production equipment were manufactured locally. The early industry was further strengthened by industry-based organizations promoting technological innovation, market competition, and export development.

Beginning in the 1930s, a series of recessions led to protectionist measures that virtually crippled the Swiss watch industry. At the time it was thought that cartelization—limiting exports of technology, parts, and know-how and protecting independent assemblers—would preserve Swiss world-market dominance. This rigid structure, which regulated the nature of technological uptake by firms, survived until the 1970s. In the meanwhile, shifting technological paradigms—most notably the advent of quartz technology and economies of scale through vertical integration—led the Swiss watch industry to the brink of disaster. Japan, and more recently Hong Kong, took advantage of Swiss inertia to launch into traditionally Swiss-dominated market segments through aggressive marketing campaigns and emphasis on technological innovation.

By the early 1970s, when foreign competition hurdled technological paradigms from mechanical to electronic watch movements, the Jura's undisputed dominance ended. Massive job loss and out-migration followed as firms, unable or unwilling to adapt to new technologies, closed their doors. Today, while world leaders in watch export value, Swiss watchmakers produce only a fraction of their pre-1970s output levels, and resources needed to invest in new product research and development remain scarce.

In a span of less than thirty years, the world's dominant watch region yielded technological leadership (in watchmaking and micromechanics) to its Far Eastern rivals. What lessons can be learned about regional economic development and technological innovation from the experience of the watch region of Switzerland?[1]

This chapter begins by elaborating the twentieth-century history of the world watch industry and the impact of technological change on Swiss industry dominance. From this historical perspective, it examines how the watch region was transformed during this period of industrial restructuring. The chapter proceeds to identify weaknesses of the region's industries and firms and to relate industry structure, technological custom, skill formation, and technological change to the process of long-run regional development. Finally the chapter concludes by examining public policies that have been enacted to assist regional recovery. This discussion suggests the degree of difficulty associated with breaking out of prior paths and highly articulated trajectories of development.

THE RISE AND FALL OF THE SWISS WATCH
INDUSTRY AS GLOBAL LEADER

Contested Terrain: The World Market for Watches

Historically, the Swiss watch industry has shown surprising resilience in the face of change. In the early years of the twentieth century, the industry was issued a major challenge by the US watch-production system. America's watch manufacturers developed machinery to produce mechanical watches at high volume with low cost, low skill, and relatively high levels of precision. Nevertheless, within this well-known mechanical watch technology paradigm, the Swiss production system generated a decisive response to American production innovations. Over the course of twenty years, the Swiss adapted aspects of the American system while maintaining their core competency in precision machining.

The early 1920s, following extreme turmoil in its traditional European markets, was a period of instability in the watch industry. Demand for Swiss watches declined precipitously between 1916 and 1921.[2] The severity of the crisis forced family businesses to take drastic steps simply to reduce inventory. Opportunism, price cutting, and increased export of movements and parts further destabilized the industry (personal communication with Luc Tissot). This unprecedented threat resulted in a call for industry regulation, and a cartel was formed.[3]

The Statut de l'Horlogerie codified the 1930s industry structure and established rules to regulate the industry. These changes resulted in a period from 1933 to 1961 during which the Swiss watch industry experienced considerable stability matched by handsome growth. But in the early 1960s three decades of relative stability once again gave way to uncertainty. Foreign competition ended both Swiss monopoly on mechanical watch production and the nation's quasi-monopoly on the world watch industry. The slow erosion of Swiss world-export market share met with cries from industry members to change the laws that had governed the industry for thirty years. In 1966 the government rescinded (effective from 1971) all remaining regulations governing the manufacture of watches. Firms were free to export, merge, buy, and sell to foreign firms. As expected, the watch industry underwent a series of unprecedented mergers. The healthier and larger establishments merged to match the sizes of their Far East Asian and American rivals. Through rationalization, restructuring, mergers, and acquisitions, the Swiss reached price parity with the Japanese.

The early 1970s were a time of profound change in the world watch industry. Free of the Statut de l'Horlogerie, Swiss manufacturers and assemblers set out to meet their major Japanese and American competitors head-on. But their production system was not easy to dismantle or rearrange. Japanese and American firms posed unique challenges to Swiss watchmakers. New and rising competition and the advent of a technological paradigm shift were both significant problems.

The Japanese industry was vertically integrated and therefore a low cost producer (Knickerbocker 1976; Maillat 1982). Japanese companies (Seiko and Citizen) made major inroads in the world watch market as both component and finished watch manufacturers. They produced high quality, low-priced movements that were sold to firms around the world. By the early 1970s Japanese watch companies had succeeded in capturing 14% of the world watch market. More than half was sold in export markets.

Government regulation and assistance (R&D grants) accelerated Japanese penetration of the world watch industry. The Japanese watch industry continued to rationalize undergoing further vertical integration that streamlined operations and reduced inefficiencies.[4] By 1978 88% of Japanese production was attributable to two firms. Switzerland's leading manufacturer accounted for only 9% of total production (*Business Month* 1988).

The most critical advantage of Japan's capital-intensive system is the ability to manufacture components in huge volumes at low cost. Large profits generated by this system have allowed Japanese firms to make dramatic improvements in process technology. Yet even this enviable position is not invulnerable to global circumstances. Low-cost competitors such as Hong Kong have cleverly appropriated the fruits of Japanese investment in R&D and movement manufacturing and used them to amass world volume leadership. In recent times, macroeconomic circumstances (particularly rising exchange rates) have also eroded Japanese market share.

The US Market and American Watch Manufacturers

Japan was not the only significant challenge to the Swiss watch industry. The United States was both the world's largest and most competitive watch market. The vast majority of American demand was satisfied by domestic firms. America's two stellar watch manufacturers—Timex and Bulova—essentially controlled two-thirds of the nation's market and presented significant problems for Japanese and Swiss manufacturers (Knickerbocker 1976; Landes 1979).[5] Both corporations followed the American system of mass production. Employing a combination of sophisticated production technology and labour flexibility (through internationalization of production), Bulova produced a range of products spanning all price categories.[6] Much of the company's high end products were produced in Switzerland.

Timex, instead, sold a product that was cheap, simplified, and standardized. Because the United States lacked skilled watch-workers,

Timex pursued a capital-intensive production strategy. Machines were automated to reduce human involvement to a minimum. The company developed highly efficient, dedicated production equipment to produce huge volumes of standardized products. Timex introduced hundreds of different models based around the original eight-step process.[7] The company designed a dramatically simplified but well-manufactured watch with a relatively long life. Given that the watch was inexpensive, Timex made no pretence of providing after-sales service. When the watch stopped running, it was generally understood to be disposable. Like Bulova, Timex established international market presence and production capacity. The company had twenty plants scattered around the globe. Each market was carefully analysed, and sales strategies were adjusted according to local customers (Knickerbocker 1976).

Technological Change and Industrial Instability

Prior to the 1970s the world watch industry grew steadily, and production was shared among three countries, the United States, Japan, and Switzerland.[8] There was an enormous expansion of markets in the 1970s when the world market for watches doubled from 230 to 450 million watches (Union Bank of Switzerland 1987). Demand was overwhelmingly weighted in mechanical devices. Only 2% of export sales were electronic watches. But in just two decades the structure of demand changed. By the late 1980s electronic products comprised 76% of world consumption—approximately 60% digital, and the remainder analogue watches.

The introduction of electronic watches in the early 1970s had a profound impact on the Swiss share of world markets (Table 6.1). In 1974 Swiss watches made up 40% of the world export market (by volume). Ten years later this figure had fallen to 10%. The loss occurred almost entirely in the high volume, low- and medium-price watch market segments.

TABLE 6.1. *Export of watch movements and completed watches, 1951–1980* (000s of units)

	Japan	Switzerland
1951	31	33,549
1955	19	33,742
1960	145	40,981
1965	4,860	53,164
1970	11,399	71,437
1975	17,017	65,798
1980[a]	68,300	50,986

[a] Includes movements.

Source: Landes (1984: 6).

With the introduction of the electronic watch, the competitive terrain shifted from precision based on mechanical know-how to accuracy based on electronic engineering. While the Swiss were the first to develop electronic watch technology, competitors succeeded in commercializing it. And although the Swiss responded relatively rapidly (within two years) to this new competitive technological threat, volume market share was permanently lost to the Japanese and, more recently, to the Hong Kong watch industry.

Swiss inability to capitalize on microelectronics technology was due to long-standing traditions governing watch manufacturing. Swiss watchmaking culture, based around mechanical watch manufacturing technology, was simply too rigid to take up the challenge presented by the new technology. Obstacles to rapid change were significant. Most importantly, the new technology was derived from a knowledge base originating outside the watch region. The Swiss lacked the technical skill to broadly innovate in the field of microelectronics and skilled labour had to be imported.

Furthermore, when quartz technology was first introduced, demand for mechanical watches was strong. Therefore market signals indicating the need for change were muted if not imperceptible. The Swiss lacked a comparable signal to push their industry toward the new technological frontier. In addition, by watchmaking standards, the time span during which a response was needed was very short. Previous innovation had been built around a two-year manufacturing cycle. Finally, efforts to embed a foreign technology into existing products were crippled by a manufacturing culture steeped in tradition. Rapid change was the antithesis of a watch culture which rewarded patient methodical actions within an existing technological paradigm.[9]

Watch-manufacturing's fragmented production structure also presented problems. Subcontracting levels were high, and the region's dominant firms could not exercise control over the myriad component producers. A fixation with precision had lulled the region's firms into complacency about their vulnerability to external forces. Supreme precision, however, did not require a theoretical understanding of new scientific developments. Rather it demanded great attention to detail. As Pierre Rossel, one close to the industry, notes, the region's firms were unprepared to overcome a technological paradigm shift that devalued the region's long-standing comparative advantage.[10]

The introduction of the electronic watch resulted in unprecedented change in the organization of watch production. Differences between electronic and mechanical watches were dramatic. Whereas labour costs constituted as much as 70% of a mechanical watch, in electronic watches labour costs were expected to be very low (less than 10%). Another major difference was the control of technology. Unlike mechanical watches, for which the Swiss effectively controlled technology, electronics technology was widely available—thus increasing the likelihood of threats from new competitors with little or no prior watchmaking experience. Given the evolution of electronics, it was almost a foregone conclusion that price decline would occur in tandem with increase in capability. Thus even the cheapest watch could be a good watch.

The advent of the electronic watch presented severe problems for Swiss watch manufacturers. Efforts required to overcome technological deficiencies associated with quartz technology required an industry-wide response. Yet given the industry's weakened condition, no single firm could afford the costs of developing such an uncertain technology. Thus numerous industry research associations formed. This new form of collaboration created serious problems because no single firm could appropriate the fruits of collective research and translate them into a competitive advantage to capture new markets. Unlike times past, when pursuit of new innovation formed the basis of market share, collective research became collective knowledge. Firms were compelled to embark upon efforts to create technological differentiation based on the original quartz innovation. These efforts were costly, uncertain, and occasionally unsuccessful.

Reorganization and Rationalization

The mid-1970s recession had devastating effects on the Swiss watch industry. World demand declined and the Swiss were particularly hard hit because the franc appreciated dramatically relative to currencies of other watch-producing countries (Katzenstein 1984; Union Bank of Switzerland 1987; Landes 1983). Starting in the late 1970s, the ensuing rapid decline in output and market share precipitated bank intervention to stabilize the Swiss industry. Hundreds of small family-run firms went out of business. Mergers occurred to gain greater access to markets. Family-run firms attempted to rationalize outdated organizational structures—creating further industry chaos.

By the early 1980s the Swiss industry was in disarray. The next recession dealt the final blow to the industry's historic organization. Faced with operating losses and massive inventories, SSIH, the major watch brand holding company, eventually became a victim of industry reorganization. The company could not solve the equation of low prices, wide assortment, small volume, rapid change, short delivery time, and wide model series (personal communication with Luc Tissot 1990). Seiko, Japan's largest watch-producer, was able to respond because it had the market volume to offer a broad assortment with economical series, low prices, and short delivery. A single statistic says it all: 'on the average Japan produced, under each brand name, 6 million watches in the 1970s compared with fewer than 100,000 in Switzerland' (Katzenstein 1984: 221).

In the early 1980s SSIH and ASUAG, the vertically integrated movement parts manufacturer that supplied the entire industry, were forced by the banks to merge. While the national significance of the Swiss watch industry could not be abandoned, either could industry organization be allowed to continue as it had in the past. The merged SMH Group was taken over by powerful Swiss industrialists. One of the most dramatic changes arising from the merger was the introduction of a wholly new product, the 'Swatch', propelling the Swiss back into the low-priced segment of the market (*Business Month* 1988). Swatch is a plastic watch manufactured at high volume using advanced automation and assembly-line methods. But the real innovation is in marketing the watch as a high fashion, mood-oriented product. Multiple ownership is stressed, and marketing is targeted toward specific age groups.

Although the Swiss share of world (volume) output continues to fall, the country remains the leading watch exporting nation by value. With only 10% of world output, the Swiss command 45% of watch export sales value. Japanese predominance in mid-priced watches is evidenced by the country's 1 to 1 ratio between volume and value of watches sold (35% of volume and 35% of world export sales). Hong Kong demonstrates the inverse of the Swiss experience, supplying 50% of world output, but commanding only 10% of watch-industry value.

The reorganization and merger of ASUAG and SSIH in the early 1980s was only a precursor to further rationalization. Industry participants cautioned that a momentary upswing in demand should not lull producers into complacency. Indeed, Tihomil Radja, chief economist of the Swiss watch-industry federation (FH), indicated increased demand was occurring in only some watch-industry segments. While strong growth was forecast for the low (primarily Swatch) and high ends of the industry, middle-range watch-producers expect heavy competition from Far Eastern rivals.[11]

THE GEOGRAPHY OF INDUSTRIAL RESTRUCTURING

The geographic impact of industrial restructuring in the watch industry has been highly spatially concentrated in Switzerland. Given the sectoral specificity of industrial decline over the last thirty years, a few cantons absorbed the majority of job loss. The country's mountainous regions experienced the most severe employment declines. In particular, half of all jobs lost in the nation since 1965 were from the Jura watch region (Elsasser 1982).

The Jura Region[12]

The industrial structure of the Jura Arc is characterized by a large number of small- and medium-sized manufacturing enterprises (SMEs). A second feature of the Jura's industrial structure is the predominance of family-run firms. While no recent census exists to buttress this statement, the history of the watch and machine-tool industries includes a heavy concentration of locally owned establishments.

The occupational structure of Jura firms differs significantly from the national average. More than half of the working-age population is employed in manufacturing (a comparable figure for the nation is only 39%). Firm surveys reveal lower salaries and fewer employees with high levels of skill in the Jura than in the country as a whole.

According to some accounts, firms in the Jura retain antiquated production processes in lieu of adopting state-of-the-art manufacturing technology. Large segments of the business community are locked into established products and markets—demonstrating little ability to innovate (Maillat 1984).[13] The dominance of manufacturing in the region has not precipitated the formation of an equally strong service sector. Massive job loss in the 1970s decade was not matched with concomitant growth of service employment within the Jura. In part this reflects the inward orientation of firms which still perform all functions in-house.[14] In fact, the effects of industrial restructuring were particularly acute in the region because service job growth was far below the national average.[15]

Machine Tools and Watches: The Economic Base of the Region

Two of Switzerland's most important industries are concentrated in the Jura—precision tools and watchmaking. These two industries each export about 95% of their products abroad—making them particularly vulnerable to fluctuations in world markets.[16]

Unlike other Swiss multinationals which shifted production to low-wage countries with growing markets, the watch and machine-tool industries have remained entrenched in the Jura region. SMH, the largest watch-manufacturer, employs fewer than 20% of its employees outside Switzerland.[17] Domestic employment concentration also characterizes machine-tool firms such as Sulzer. Jura firms continue to automate routine production instead of outsourcing and will undoubtedly undergo further industry shrinkage. Automation and firm closures have resulted in massive job losses which have led to a striking population decline.

Industrial employment in the Jura fell by 41% between 1970 and 1980. Most industries in the region underwent contraction, but none so dramatically as watch-making. Nationally, watch-industry employment plummeted from 89,450 in 1970 to 32,000 in 1985, an alarming decrease of 64% as shown in Table 6.2. A similar level of job loss occurred in the Jura. Table 6.3 illustrates the extraordinary employment losses in the watch, engineering, and metal industries between 1970 and 1984.

Employment statistics do not paint an entirely accurate picture of the

TABLE 6.2. *Changes in companies and jobs in the Swiss watch industry, 1970–1985*

	No. of establishments	Jobs	Jobs per establishment
1970	1,620	89,450	55
1973	1,260	75,800	60
1976	1,080	55,200	51
1979	870	46,700	54
1982	730	38,200	52
1985	600	32,000	53

Source: Annual Census of Employers Convention, as cited by Union Bank of Switzerland (1986).

TABLE 6.3. *Industrial employment in the Jura region, 1970–1984*

	Total	Metals	Machines	Watches	Others
1970	95,120	9,213	27,145	44,999	13,763
1975	72,148	6,995	21,697	32,243	11,213
1980	64,933	6,869	18,796	26,201	13,067
1981	64,435	6,971	21,569	25,040	10,855
1982	63,739	7,565	20,940	23,030	12,204
1983	60,964	8,194	18,623	20,492	13,655
1984	56,023	7,468	18,938	17,643	11,874

Source: Vasserot (1986a: 3).

Jura's population loss for two reasons. Foreign workers—among the first to be let go in times of economic depression—are effectively not counted, and changes in the labour participation rate further reduce the base of individuals who might otherwise be considered unemployed. In 1986 25% of the Swiss labour force consisted of foreign high- and low-skilled workers. Between 1974 and 1977 the number of foreign workers in Neuchâtel Canton alone dropped by over 50% in response to the crisis in the watch industry.[18]

Within the Jura Arc, job and establishment decline were unevenly distributed among cantons. Job loss was particularly high in the outlying areas of Solothurn, Ticino, and Vaud. In and around Geneva, job loss was less severe. This uneven pattern relates to the geographic dispersion of firms specializing in watch market segments. Plants producing low-cost watches were most severely affected by competition from low-wage countries. Cantons where low-cost watch manufacturing was concentrated experienced the greatest employment and establishment declines (Table 6.4).

TABLE 6.4. *Reduction in watch-industry companies and jobs by canton, 1984*

Cantons	Companies	Workforce
Geneva	59	83
Vaud	59	31
Neuchâtel	42	45
Basle region	41	48
Berne	35	33
Ticino	34	28
Solothurn	30	31
Other cantons	46	39
AVERAGE	·39	39

Note: 1970 index = 100.
Source: Union Bank of Switzerland (1986).

TRANSFORMING THE REGION: THE DIFFICULT TRANSITION OF CULTURE AND INSTITUTIONS TOWARD A COMPETITIVE FUTURE

The literature on flexible districts emphasizes the key role played by culture and institutions in the evolution and transformation of regional production systems. The case of the Swiss watch industry and the Jura region is no exception. In the industrial district literature, institutions are seen as regulatory mechanisms that help keep the system flexible. But how flexible are these institutions? In emphasizing the flexibility of institutions I am concerned about firms' abilities to adjust their core competencies to make effective use of new technologies. This section illustrates the liabilities associated with sectoral dominance of a region's economic base. It also shows how complex-based relationships and rigidities shape firms' abilities to break from prior industrial traditions and to pursue new inter-industry capabilities including product design, the incorporation of advanced services into existing products, and the use and development of new intermediary roles. The section examines how these institutions and institutional relationships are evolving during a period of heightened international competition and technical change.

The Industrial Structure of the Jura Arc: Shifting Paradigms?

The bulk of empirical work on firms in the Jura[19] attempts to discern whether the region's establishments (and therefore production system) are changing from a paradigm based on mechanical technology toward a multi-sector economy based around information technology. Indeed, early research describes the Jura watchmaking region as largely mono-industrial, dominated by mechanical watch manufacturing (Maillat 1982). With the exception of the machine-tool industry, which largely grew out of watchmaking, no other industries established a significant presence in the region. And as one scholar (Landes 1984: 48) noted: 'the dominating position of watchmaking . . . had a dampening effect on business enterprises in other sectors'. Swiss industry maintained its geographic dominance until foreign competition eroded the absolute comparative advantage derived from the decentralized production structure. 'Whereas the [foreign competition] was vertically integrated, the organization of Swiss watch manufacture relied on a very slow transition from horizontal to vertical activities, starting upstream and thus neglecting the highly vulnerable downstream structure' (ibid.).

Speculation exists about whether application of microelectronics to industries in the region constitutes a successful transition from a mechanical to an electronic paradigm. Divergent views abound. Maillat (1982: 18) suggests 'the electronic watch forced the gates of the region open'. Another view, however, suggests the two paradigms appear to coexist: 'Although the [traditional] production system . . . has been partly dismembered, it must be recognized, nevertheless, that two new partial "industrial substructures" have emerged' (Hainard 1988: 2). This implies that new technology has not really changed the existing institutional context. Indeed, with the exception of the largest firms, production technology remains the same.

The experience of the machine-tool industry illustrates the difficulties that firms face as they attempt to incorporate new technologies into existing products. Interviews with both major and minor machine-tool companies suggest that the prior precision-based industrial paradigm shapes firms' responses to new competition and circumscribes acceptable design modifications made to existing machine-tool product lines. In other words, firms' applications of microelectronics to machine tools reflects incremental adjustments to existing product designs. Failing to commit to radical departures toward a new industrial trajectory, few attempts have been made to move away from single purpose machine designs and toward multi-purpose product development. For example, the largest machine company in the region maintains its adherence to precision (and over-engineering) in the face of significant cost pressure from Far East competitors. When confronted with a 40% cost differential between its products and those of its Japanese rivals, the plant manager replied, 'the cost differential can be explained by the fact that our machines remain true (to cutting) for the life of the machine (about 20 years). Our competitors cannot make a similar claim.' Unfortunately, the rate of technological progress in the machine-tool segment (in which the Swiss firm competes) is so rapid that product obsolescence occurs long before the machine's lifespan has been exhausted. Japanese competitors have reacted by scaling back the longevity of their machines while adding new features to make their machines more flexible.

Breaking with an industrial tradition demands more than reconfiguring product designs. In the case of machine tools, competing globally requires the transformation of everything from a machine-tool company's culture—the identity of the firm as seen by its owners and the workers in the factory, to the firm-based industry-wide apprenticeship programme which relies heavily on learning-by-doing to instil the ethic of precision in workers.

Few Firms Innovate

The bulk of research on innovation, adoption, and development in Jura firms suggests that firms are engaged in incremental industrial innovation in the spirit of what Dosi calls 'natural technical progress within an existing paradigm' (1982: 154). Most new developments are modifications to existing products and processes and cannot be viewed as true innovation. Firms rarely plan to pursue new innovations. Rather, innovations have occurred as firms have reacted to market conditions.

A substantial effort has been made to determine the degree of technology diffusion in Jura firms. This research also gauges the significance of the innovation process and the degree to which local firms are innovative. These studies identify the manner in which new technology enters the region, and how differences in regional supply factors affect firms' ability to innovate.

In one study, Boulianne (1982) found that slightly more than one-third of firms interviewed in the region had introduced either a new product or process. Further probing indicated that these adopting establishments were larger, more mature (between thirty and seventy years old), more capital-intensive on average, and were not primarily from the watch industry (1982: 49).[20]

In the diffusion process, information carriers were technical personnel and new machines purchased from other firms. Dynamic firms with high rates of innovation adoption employed large numbers of workers in tertiary tasks. They also purchased sophisticated business services outside the firm.[21]

In a large survey project conducted in 1985 Maillat and Vasserot identified firm characteristics associated with new product and process innovations (Maillat 1986; Vasserot 1986*a* and *b*). Of the innovating firms, the majority introduced innovations which represented minor modifications discovered by chance rather than derived strategically. Only a fraction of all firms (less than 12%) originated radical innovations. Examination of the evidence indicates that small firms innovate at expected levels given their representation in the study data. While the application of general automation was reported by a fair number of firms, this is not a new process innovation *per se*. No surveyed firms created wholly unique process technology.[22]

Firms that introduced both process and product innovations were larger on average and engaged in strategic planning for new developments.[23] However, these firms did not boast a technologically sophisticated labour force (Maillat 1986: 18).

Creating New Industry based on Regional Core Skills

Redeploying a region's core skill competencies is made difficult by cultural inertia and individual self-interest. Resistance to the manufacture of new products is often coupled with the existence of technological bottlenecks. In the first section I highlighted the difficulty that Swiss firms had in recognizing the significance of a new technology, particularly when an innovation broke decisively with tradition. One highly illustrative example of this tendency is the case of the Swiss innovator Hetzel's invention of tuning-fork technology. Swiss watchmakers resisted in adopting the new technology, and Hetzel turned to American Bulova to market his find. Many other innovations that have been created in the region have met with similar resistance. As a sales representative of the micromechanical tool division of SMH noted, 'we developed many new products with real market potential. We just never received the support we needed and could not control pricing policy in order to bring a new product to market competitively.'

For decades the Jura mountain region's competitive advantage lay with its ability to manufacture products with high levels of precision and great detail. Although the region's watch industry experienced significant decline, the core skills of precision machining remain in the region. In the early 1980s a local entrepreneur undertook an experiment to identify industrial activities that would reuse the region's core skill.[24] After investigating numerous products, he introduced the manufacture of coronary pacemakers into the region. The production process bore remarkable resemblance to watch-manufacturing, and employees found the work satisfying and rewarding. Whereas a watchmaker's previous task had been to keep the world on time, pacemaker manufacturing had an even greater responsibility—keeping the human heart regulated.

Entrepreneurs continue to search for new products compatible with the region's core skills. The original experiment with medical equipment illustrates how a region can reuse long-established core skills. And although it remains difficult to exploit the region's skills (in part due to a lack of complementary business services and resistance to change), the pacemaker experience highlights the value of matching a region's existing industrial culture and tradition with new technologies and products.

Regional Institutional Weaknesses

A synthesis by Crevoisier *et al.* (1989) merges many insights from previous work into a useful summary of the region's institutional and cultural weaknesses. The authors emphasize the weakness of the mountain region's training system. While firms exhibit empirical know-how—the practical relationship to the production process—they note an absence of 'analytical know-how: the additional existence of an ability to carry-out a relatively-thorough scientific analysis of the factors involved in that process' (1989: 3–4).

Other scholars contend that the Jura has been primarily a site of high-precision assembly. While residents of the region have historically made significant contributions to watch-manufacturing technology, the knowledge base was relatively small and controlled by a few individuals. François Hainard (1988) argues that the Jurassien style is associated with docility, dexterity, and skill at working on small pieces. This ability is more pronounced in the Jura than in other Swiss regions or in neighbouring countries. He argues that few members of the region—entrepreneurs or even a few élites—ever had the ability to conceptualize new products and create new technologies. In contrast with conceptual knowledge, Hainard feels that the 'ability to do' (*savoir-faire*) is not synonymous with dynamism and innovation. On the contrary, it can lead to complacency, recalcitrance, and production delays. In essence, the Jura's critical competitive advantage was precision assembly. The region generally lacked the intellectual capacity to innovate autonomously. He suggests that the region's problem is perceptual, arising from long-standing dependence on a single industry.[25]

The Jura region boasts a number of institutions providing technical information and advanced production acumen. These include high tech research centres such as the Centre Suisse d'Electronique et de Micro-technique (CSEM). But in general, research organizations have difficulty filtering scientific knowledge to firms in the immediate vicinity. In part this is because firms receive technical information through different channels from those created by the scientific laboratories. More fundamentally, this type of basic knowledge is most readily absorbed by large firms that have research and development departments capable of using theoretical breakthroughs in technology.

While there is electronic component production capacity in the region, some observers believe these producers operate at suboptimal levels and lack the ability to create significant technological developments with potential for downstream commercialization. Moreover, because the intended market for these electronic products is the watch industry, the technology is based on the characteristics of the end user. This makes it doubtful whether the region can develop an independent microelectronic base separable from watchmaking.

A classic agent of change referred to in the literature on local industrial complexes is educational institutions that are thought to be reservoirs of new knowledge. While the Jura like many complexes based around craft skills enjoys a number of training institutes and technical research facilities, the region still has an acute shortage of technically qualified personnel. Despite the presence of such technical centres as CSEM, most educational facilities in the region are still oriented toward the vocational skills formerly tied to watch-manufacturing. And in watchmaking and machine-tooling in particular, the educational emphasis has traditionally been on apprenticeships or on-the-job training (as opposed to research-oriented education).

Recruitment of skilled labour is a critical problem for firms in the region. Branch plants have had to hire skilled workers from outside the region to operate their factories. Local manufacturers are also increasingly turning to foreign workers to provide technical skills.

Regional firms are experimenting with programmes to resolve skilled labour shortages. Several machine-tool firms have joined together to develop a firm-certified curriculum. Enrolment in the programme leads to an industry-approved certificate. Upon completion, a graduate is considered qualified to work in all participating firms. Workers can move between firms in search of a satisfying working environment.

Watch-industry officials are also concerned about the problem of skill shortages. One official indicated in an interview that the industry needed to reinvigorate the profession via a publicity campaign designed to attract young people back to the field and the region.

THE SEARCH FOR RESPONSES TO GLOBAL COMPETITIVE PROBLEMS

On the eve of the 1990s the Jura Arc found itself increasingly influenced by international economic events. While historically the region's production system has been internally integrated and relatively autonomous, the question today becomes one of how the region's firms choose to participate in a global system of production and consumption.

The following discussion suggests that Jura firms are selecting short-term paths of integration by pursuing strategies that incorporate notions of static flexibility. To remain price-competitive with lower cost countries, Jura firms are increasingly relying on subcontracting, automation, and de-skilling to reduce costs. Ironically, this strategy, which allows for immediate industry survival, may preclude long-term autonomy and innovative ability. Like firms in other regions of industrialized nations, firms in the Jura must address the problem of how they can compete on the basis of both cost *and* skill—a situation successfully confronted at different points in the industry's history. The following discussion highlights recent developments in the Jura and raises questions about the need for a strategy which emphasizes a region's long-standing core skills during a period of rapid competitive and technological change.

Initial Forays into the Global System of Production[26]

It is important to begin by referring here to the complex's early reaction to international competition. This review illustrates how prior institutional tendencies shaped Jura firms' initial forays into the global production system. The examination simply underscores how culture can act as a *limiting* factor in the pursuit of factors to shore up profitability.

Given stiff price competition from all sides, in the late 1960s the Swiss industry made efforts to relocate production abroad. For example, Swiss Time Hong Kong (EST Hong Kong) began outsourcing watch-cases and watch-bracelets in 1969 through Economic Swiss Time Holding, a parent organization. In 1971 SSIH bought *this* company and regrouped it. By 1978 Economic Swiss Time Holding was incorporated into EST Mumpf in the canton of Argovie, EST New York, and EST Hong Kong. Rationalization ensued, and each part of the parent company was either sold to a private concern or dismantled.

Mondaine Watch was one of the major Swiss watch groups in the 1970s (fifth in 1974 after ASUAG, SSIH, SGT, and Rolex). In 1974 it was transformed into the Asian Swiss Industrial Company (ASICO). From the beginning, this enterprise was assembly-oriented, and in the late 1970s it expanded assembly activities into Brazil. Changes in the low end of the watch industry in the early 1980s precipitated reorganization of the company, resulting in a much smaller workforce and a shift toward electronics.

There are additional examples of Swiss watch companies that attempted outsourcing in other countries during the 1970s and 1980s. These include Baumgartner Frères SA (BFG), Ronda SA, and Beltime Watch Company of Switzerland (Hong Kong) Ltd. In fact, all the major Swiss watch groups attempted foreign production or assembly at one time or another. However, in each case the decision to begin outsourcing was made on a piecemeal basis—without a joint Swiss industrial strategy. In response to difficulties with new technologies such as quartz (and later to their lack of flexible response), each group attempted independently to transfer production abroad.

Swiss transfers of manufacturing proved unsuccessful for various reasons. When Swiss companies finally internationalized production, they were disadvantaged due to rising exchange rates, the oil crisis, and foreign competition. By the early 1970s capital, technology, and business skills were in short supply. Furthermore, watch companies had little experience managing foreign operations. Swiss manufacturers lacked the required organizational flexibility and cultural sensitivity to operate production facilities in different countries. For example, manufacturers tried unsuccessfully to implant their own hierarchically organized production system into Hong Kong instead of organizing production in accordance with Hong Kong's manufacturing practices. High turnover and low productivity plagued the Swiss production enclave and ultimately forced companies to abandon their foreign operations (personal communication with Luc Tissot 1990; Blanc 1988). The benefits Swiss companies hoped to enjoy by shifting production to Hong Kong—low cost, rapid change, flexible labour—eluded them. Cultural myopia inhibited Swiss firms from successfully internationalizing production.

Swiss watch companies' initial attempts to create a global division of labour failed largely due to existing patterns of business behaviour that proved ineffective outside the Swiss production context. Swiss producers' second attempt in the early 1980s to insert themselves into the global division of labour has been far more successful. Nevertheless, greater success has come at a high cost to the region and to the social fabric of the industry.

The Emergence of Swatch

As noted in previous sections, with the industry on the brink of bankruptcy, the state asserted fiduciary control and transferred ownership to individuals who had no prior history in the region or in the industry. In the early 1980s the SMH Company, which was given complete licence to reconstitute large segments of the industry, created an entirely new production system that promotes an industry image based on fashion and operates the industry on the basis of mass manufacturing practices, subcontracting, and new distribution strategies.

With the introduction of the Swatch watch, SMH moved the industry away from its age-old watch marketing concept based around precision and toward a new fashion-based strategy. This reconceptualization led the Swiss industry back into the low-end price segment they had abandoned to Hong Kong and Japanese competitors.

This shift precipitated dramatic changes in the structure of the social relations and production system within the complex. The major innovation was the creation of a vertically integrated, highly automated continuous process manufacturing system to produce Swatch watch parts and to execute final assembly. To accomplish this industry reconfiguration the watch calibre (dimension) was standardized, and the number of parts in the watch was drastically reduced. Product differentiation is now accomplished by varying the exterior colour and (within limits) material of the watch, and altering the watch face to reflect different product moods or messages.

Simultaneous with the introduction of Swatch, SMH undertook a major reorganization and rationalization of mid-price branded products (formerly owned by the SSIH Group) and gained further control over the industry reorganization. This phase of adjustment also included introduction of new models such as the Rock Watch. The use of new materials was a significant deviation from past practices when watches were manufactured almost exclusively from metal. As part of this reorganization, new exterior parts inputs (cases, bands, etc.) have been sourced from all over the world, and interior movement parts manufacturing has been standardized. As a result, mid-range watch movements have converged technologically. Yet achieving economies of scale in parts manufacturing has resulted in the erosion of technological distinctions between brands. Brand differentiation is now almost exclusively based on characteristics associated with the exterior of the watch (metal finish, jewelling).

The Internationalization of Watch-Manufacturing

Multinational corporate solutions to the Jura's competitive problem are precipitating dramatic regional change. In February 1988 ETA, the largest Swiss watch conglomerate (and part of the SMH Company) closed several factories in the Jura in an effort to compete more efficiently in the world market. Under pressure from the Swiss metalworkers' union (FTMH), the local government in Saint-Imier made demands to ETA to substitute a production unit equal to the one that was closing. While public dissent has had little effect on the rate of plant closures, the latest wave of closures has met with stiff resistance from local residents who complain that such actions are further eroding the industry's prominence in the region.

Another example of labour's growing voice in regional policy concerns shift-work in factories. Although the Jura has long been the site of assembly operations, there have never been widespread twenty-four-hour factory shifts. Swatch's success can be attributed, in part, to labour intensification. Under pressure from the labour unions, the Swiss government initially declared this practice unconstitutional. And although shift-work has been legitimized, Swatch's comparatively lower wages and longer working hours are still under fire by employees.

Clearly a low-wage competitive strategy does not sit well with the majority of the Jura's workers. And Jura firms cannot realistically expect to compete with Asian rivals on a cost basis alone. Multinational firms recognize that they are competing with countries where wages are not only lower, but workers are better trained. Regions like the Jura must reorient away from cost-minimizing toward profit- and skill-maximizing strategies (Blanc 1988).

The emergence of Swatch demonstrates the adaptability of the region to significant exogenously induced change. It is ironic that partial revitalization of the region's industry, based around the plastic watch, reflected material competencies that substantially pre-dated the crisis. In the late 1960s the Tissot Watch company developed and mass-produced a gear-driven plastic watch. Although several hundred thousand watches were produced, the innovation ultimately failed to achieve widespread market acceptance for many of the same reasons that microelectronics stumbled in the recent period. Industry organization in the 1960s exacerbated competition among the major companies over whose brand name would distribute and market the plastic watch. Because no agreement could be reached, the potentially lucrative, and clearly innovative, product was abandoned.

The Future

Researchers point out differing paths to regional economic recovery. Vasserot calls for a stronger network of SMEs and more significant inter-firm collaboration and research–enterprise links, Maillat emphasizes closer linkages between manufacturing and tertiary activities. He finds the 'traditional divisions between the secondary and tertiary sectors have been replaced by new structures in which scientific knowledge and production processes are combined at the strategy-development level and the manufacturing level itself'. Maillat advocates a regional production system based around a 'production filiere' comprising SMEs supplemented by upstream tertiary activities (data collection, feasibility studies, design, and development of products), downstream activities (market surveys, sub-contracting, advertising), and horizontal management support services (Maillat 1989: 19).

Finally, the recent research by Maillat (1989) and others focuses on production strategies and the application of flexible technology to create niche market products. I would contend that a focus on production strategies draws industrial policy discussion boundaries too narrowly. For example, flexibility is only useful in conjunction with a set of core skills that can be successfully deployed and redeployed in the face of market uncertainty.

INSTITUTIONAL RESPONSES IN THE JURA

In this section I briefly review the public-sector response to regional restructuring. It will be suggested that the public sector remains largely reactive in the face of technological change and global competitive pressures. Government response to industrial restructuring has taken the form of policy development designed to revitalize declining regions. Programmes largely respond to needs of the prior industrial order and are thus limited in their ability to sow the seeds of regional change. The two major federal-level programmes are the LIM (see below) to aid mountain regions and the Arrêté Bonny to encourage innovation and diversification in threatened economic regions. Local initiatives in the Jura and Neuchâtel cantons also encourage economic development. Swiss policies show little variation from standard practice in other developed nations. Emphasis is placed on small- and medium-sized business, yet there appears to be little concern about the appropriateness of such a focus and the effective scope of resulting policy initiatives.

Two federal programmes target the Jura specifically. One of these is La Loi sur l'Aide aux Investissements dans les Régions de Montagne (LIM).[27] The thrust of this federal aid programme is to improve living conditions in the mountain regions in order to stem population out-migration. To qualify, an area must have a regional development programme recognized by the federal government, and cantons must match federal funding and execute the programme for which they receive funds. Aid consists of guaranteed loans with favourable terms and is granted largely for development of infrastructure.[28] Because of the structure and distribution of firms in the remote regions of the country, government policy favours small- and medium-sized businesses. Emphasis is placed on picking firms that already exhibit success rather than establishing a broad basis for local economic development. Such policy inadvertently supports maintenance of the status quo and is relatively risk averse.[29]

Another large federal programme is the Arrêté Bonny—legislation for regions with endangered economies.[30] In order to qualify, a region's economic base must be concentrated in a single industry which is experiencing severe employment reduction. Financial aid is distributed through investment credits for purchase of machines, equipment, licences, buildings, and new construction. The federal government bestows loans initially, and the cantons must insure them.[31] By June 1985, of 200 projects funded, the Jura region received the lion's share—173.[32] Projects have been primarily in the fields of electronics, electromechanics, and machine construction.

Arrêté Bonny has had positive impacts on the Jura and Neuchâtel regions. None the less, the programme has problems. One of the main criticisms is that grants are given only to projects sufficiently developed to present the minimal risk. Thus aid is not supporting the research and development activities of firms—a chief target of the programme.

In addition, in the late 1970s cantons began enacting legislation aimed at economic development. Neuchâtel has the most innovative economic development programme in Switzerland. The 'Neuchâtel Way' targets entrepreneurs in both regional and foreign firms. Economic promotion is presented as a complete package of allowances and grants to assist prospective firms in a move into Neuchâtel. Regional promotion focuses on the area's long-standing core skill—precision manufacturing at small tolerances—and the availability of a disciplined, relatively low-cost labour force.

Quantitative information documenting the effects of Jura economic development policy is scarce (see Corat 1985). Although some data on programme outcomes exists, reporting numbers of jobs, levels of incentives, etc., the results are difficult to interpret. In part this lack of information stems from the relative newness of many of the economic development programmes. Yet given the large sums of money spent on economic development programmes and research, it seems vital that there be some evaluation of programme output to determine future policy orientation.

The Swiss system of local, cantonal, and federal economic development programmes mirrors that found in other developed countries. Tax holidays, worker training, wage subsidies, infrastructure, etc. are all features of traditional economic development activities.

Successful economic development programmes have been those which provide training and infrastructure. Firms readily admit that these two programmes are important determinants of their location decisions and firms are therefore often willing to share costs of human capital development-related activities. As with tax incentives, communities must increasingly offer these programmes just to remain on par with their competitors. Yet these programmes improve the overall condition of the community.

Public policies designed to initiate change often lack the mandate to transform a region. The corporatist system in Switzerland, based on limited state involvement in the economy, meant that the powers of the state could be invoked only on the eve of crisis. The sweeping change that occurred as the region's dominant watch companies were merged into SMH was unprecedented. But in most instances the state's role has been restricted to making adjustments to the existing system rather than altering the trajectory of the local industrial complex.

CONCLUSION

Previous discussion suggests that regional fortunes are increasingly intertwined with global events that are largely beyond a single community's control. External economic forces influence the trajectory of a region's development through their effects on locally established firms. Relationships between regions and industry are therefore reciprocal. As an industrial culture comes to define a place, it structures future opportunities for innovation. Given that job creation tends to mirror a region's skill base, and supply constraints govern the availability of mobile technically trained labour, policy has only modest effect on re-establishing regional innovation.

The long tradition of watchmaking and precision machining is the technological base for the region, yet this history is also responsible for many limitations to development. The intent of this chapter has been to illuminate the barriers to future regional innovation. The Swiss experience must be understood within a framework that exposes how technological paradigm shifts challenge previous ways of organizing an industry, culture, and society. Paradigm shifts present a series of strategic turning points that industrial leaders must navigate during a period of technological change. The Swiss are no exception. The industry's reaction to technological unknowns, given a fragmented production system, institutional inertia, and significant sunk costs in capital and equipment, were rather common responses of firms and industries undergoing radical technological change.

A paradigm shift represents an extreme event that can precipitate disjunctions between past and future means of regional development. It is clearly important to distinguish between adjustment costs associated with incremental versus discontinuous change. As this example illustrates, these two types of environmental change often converge. In the face of first incremental and then discontinuous change, a system's reaction may be dulled by market signals that mask the need for radical transformation. The ability to make incremental adjustments often creates a false sense of flexibility and therefore lulls complex members into believing that they are invincible. As this example illustrates, a complex's pre-eminence is often endogenously derived. Inevitably, internal actors may have difficulty remaining open and sensitive to external signals, thus threatening the viability of the region as an internally integrated industrial complex. And inability to change may be made manifest in a condition not unlike that of the proverbial ostrich with its head hidden in the sand.

NOTES

This research was generously funded by the Tissot Economic Foundation. The research could not have been completed without the selfless efforts of staff of the Tissot Foundation including Pierre Rossel, Jocelyn Tissot, and Luc Tissot. Special thanks goes to Ms Wendy Tiallard who graciously travelled with me throughout the region during interviews with industry representatives. She also provided exceptional help in accumulating the necessary documents to complete this work. In addition, industry representatives and former FH members, in particular Mr Rene Retornaz, greatly informed this research. Members of the Watch- and Tool-makers union and former watch-workers were of invaluable assistance in understanding the relationship of the working person in the complex. Scholars of the region, in particular the faculty at the University of Neuchâtel, in particular D. Maillat, created an intellectual road map that made the understanding of regional institutions and their ability to change comprehensible. Bettina Brunner provided excellent research and conceptual assistance not to mention flawless translation skills that greatly facilitated the development of the history of the industry and the region. Discussion with colleagues including Ash Amin, Mike Conroy, Meric Gertler, Bennett Harrison, Amy Kays, Erica Schoenberger, and others was helpful in bringing this work to fruition. Amy Kays provided excellent editorial assistance. Obviously, any and all mistakes are solely attributed to the author.

1. Developments in the international economy—technological change, global competition, corporate reorganization, and state-led development strategies— are increasingly important determinants of subnational development. Given rapid technological advancements, theories of regional development based on comparative advantage and regional factor endowments are becoming obsolete. The experience of newly industrializing countries (NICs) illustrates that nations can create comparative advantage through state-led development planning. Many industrialized regions now directly compete with these newly industrializing countries where lower costs and relatively skilled labour combine to increase technological advances. And technological competition is leading to the development of homegrown technologies—a remarkable change from the past.
2. Disruptions in the watch market presented the Swiss with new and different problems (Knickerbocker 1976).
3. It was the larger firms which had made capital investments in equipment that requested interventions designed to inject order into the historically anarchistic industry. They had to be able to control the small firms that easily sprang up to produce cheap watches.
4. Post-war Japanese industrial organization consisted of large business groups and trading companies. Since the 1960s the Japanese manufacturing system has experienced profound rationalization. The existing system has evolved from high levels of vertical integration to the present situation of high levels of subcontracting (Glasmeier and Sugiura 1991).
5. American watchmaking firms were dominant in the United States partly because of high tariffs implemented to protect the domestic industry.
6. The firm became the industry leader by scattering production around the world and trading off wage levels for labour skill. As part of this strategy, the company also subcontracted with Japanese component producers for low-priced movements (Knickerbocker 1976; Landes 1983).
7. Timex also developed true interchangeability. Parts could be exchanged not only within but between plants (Landes 1984).
8. We do not include Eastern Bloc countries' production in these figures. A considerable volume of watches is produced in the Soviet Union, East Germany, and other Eastern Bloc nations (Knickerbocker 1976).
9. A working research note by Pierre Rossel on the Jura region suggests that a convergence of factors limited the industry's ability to respond to this external shock. First, rising oil prices and currency revaluations eroded the price competitiveness of Swiss products. Thus on the eve of the microelectronics revolution, firms were already weakened by escalating costs. The early 1970s also witnessed the end of an era in which watch manufacturing had been

dominated by individual families. Many manufacturers were reaching retirement age just as quartz technology was being introduced. Producers lacked the will and the resources to take up the challenge presented by the new technology.

10. It should be pointed out that Timex and Bulova were caught in the same predicament as the Swiss. After the introduction of the quartz watch, Bulova no longer had an exclusive claim that the company produced the world's most accurate watch. Quartz watches were accurate and increasingly inexpensive. Timex was also caught by the introduction of quartz. The company did not initially have quartz watchmaking capacity, and its eight-step production process was made largely obsolete by the commercialization of quartz.

11. As one watchmaker put it in an interview, 'we [producers] at the high end of the industry were largely unaffected by the 1982 recession. Because we face inelastic demand, our limited clientele continue to buy—regardless of economic circumstances. Those firms in the middle range, however, were severely affected. Now enjoying strong demand, some firms are once again becoming complacent. This could be dangerous because the market is so unpredictable.'

12. The Jura Arc, which lies in the north-western part of Switzerland, consists of two sub-areas: the mountainous chain and the foothills of the Jura. The Jura straddles the Franco-Swiss border and includes the cantons (districts) of Jura, Neuchâtel, Jura Bernois, and Nord Vaudois.

13. e.g. while the global machine-tool industry gravitates toward multi-function machining centres, Jura machine-tool firms manufacture single-purpose equipment employed in high-volume mass-production systems. And while there are select establishments using robotics and laser technologies, there is still little technology transfer among regional firms.

14. Because firms do not contract out for higher order services, the region's share of service firms is lower than the national average.

15. Between 1965 and 1975 the Jura lost 3,000 jobs per year while gaining only an average of 600 jobs per year in the service sector (Elsasser 1982).

16. SMH, the major watch-holding company in Switzerland, is located in Neuchâtel and manufactures key watch brands including Swatch, Longines, Rado, Tissot, and Omega.

17. As part of the reorganization agreement, SMH shareholders were required to maintain employment in Switzerland and were prohibited either from selling the firm before 1992 or from selling it to foreign interests.

18. Changes in labour-force participation rates further masked the statistical impact of employment decline in the region's dominant industries. In 1980 almost 30% of Switzerland's population was over the age of 65. During the 1970s, the worst period of employment decline, many workers were at or near retirement age. These workers simply chose to retire rather than look for alternative work. Home work was essentially eliminated, and women left the factories but remained in the region with their families.

19. At the outset it is important to indicate that much of the English-language literature on the Jura region does not establish baseline characteristics of empirical studies. In many of the analyses it was impossible to determine, for example, whether the studies reflected a statistically representative view of the

region's firms. Most empirical reports do not adequately describe the sample interviewed relative to the overall industrial distribution in the region. Nor is there information about the response rates or characteristics of non-respondents. Therefore this analysis should not be viewed as definitive but rather suggestive of possible firm-level characteristics and development potential.

20. These establishments were neither the fastest growing nor the largest job generators in the region. Rather they emphasized depth in specific technology areas and redefined positions rather than adding new jobs as technology changed. Furthermore, they were not located in the most remote areas of the mountain region.

21. The studies noted that acquisition of advanced technology occurred as new employees were hired from outside. These same organizations followed strategic plans when undertaking new product development (Boulianne 1982: 60; Brugger and Stuckey 1986). They found that information transfer occurred through pre-established channels of communication.

22. Process innovators were subcontractors in the metal trades industries and ranged in size from very small to medium-sized. Very few of these firms followed a preconceived plan when adopting new technology.

23. The introduction of new process innovations followed similar lines. The most widely introduced technology was numerical control equipment. Given that computer numerical control has been in use for twenty years, this result is somewhat ironic (Kelley and Brooks 1988). Very few firms used robots, process control systems, or lasers—three new technologies used in machining operations.

24. Luc Tissot, president of the Tissot Economic Foundation.

25. Comments about the Swiss machine-tool industry by Professor Prouvost parallel Hainard's conclusions. Swiss industries, particularly small and medium-sized businesses, are locked in a technological paradigm which emphasizes precision. Interviews with machine-tool firms indicated that the firms viewed their key competitive advantage as precision metal machining and assembly. But while this is a useful core skill, it is by no means unique to the Swiss. Anecdotal evidence supports the contention that in lieu of competing at the cutting edge of technology, firms hide behind this label. As Prouvost noted, Swiss machine-toolers receive orders from Japanese firms because Japanese machine-tool companies have stopped making single-purpose equipment. Thus while Jura-based machine tools are still being made and sold, they are not at the cutting edge of technology, and their design is driven by customer request rather than direction from Swiss machine-tool companies themselves. Over the long run, retreat to niche marketing implies that machine-tool makers cannot compete on a high volume basis.

26. This section was drawn from M. Blanc's (1988) book on the Swiss watch industry.

27. LIM legislation was enacted in 1974 and modified in 1984.

28. In some instances the government will assume interest costs for infrastructure development in regions.

29. But the LIM programme has met with limited success. Because the programme was implemented during a period of economic crisis, it is difficult to determine

total benefits. In 1984 additional funds (500 m francs) were added to the LIM programme. By 1985, 2,125 projects had received funding, and aid totalled 592 m francs.

30. Enacted in 1978 (and modified in 1984), the law promotes innovative projects with potential for diversification in regions undergoing difficulties.
31. Banks provide market-rate loans to companies qualifying for support. Individual firms are responsible for innovation and diversification. The Arrêté Bonny was modified in 1984 to be more flexible and to target technology firms. Although originally intended as a stop-gap measure, the programme was extended to 1994.
32. 200 industrial projects were sponsored through the Arrêté legislation. A total of 780 m francs were spent. Of this, 204 m francs were jointly provided by the federal government and the cantons. Small and medium enterprises (SMEs) benefited most from the legislation.

REFERENCES

Blanc, J. F. (1988), *Suisse–Hong Kong: Le défi horloger* (Geneva: La Collection Nord-Sud de l'Institut Universitaire d'Études de Développement de Genève).

Borner, S. (1986), *Internationalization of Industry* (Zurich: Springer-Verlag).

—— (1989), 'Swiss Competitiveness: Where Do We Stand and Where Are We Heading', mimeo, Federal Institute of Technology, Zurich.

—— Stuckey, B., Wehrle, F., and Burgener, B. (1985), 'Global Structural Change and International Competition Among Industrial Firms: The Case of Switzerland', *Kyklos*, 38/1: 77–103.

Boulianne, L. (1982), 'Technical change: Firm and region, a case study', in Maillat (1982).

Brugger, E., and Stuckey, B. (1986), 'Regional Economic Structure and Innovative Behaviour in Switzerland', *Regional Studies*, 21/3: 241–54.

Business Month (1988), 'Up from Swatch' (Mar.), 57.

Corat, P. (1985), *L'Incidence de la politique industrielle régionale: Le cas des régions dont l'économie est menacée* (Neuchâtel: Institut de Recherches Économiques et Régionales).

Crevoisier, O., Fragomichelakis, M., Hainard, F., and Maillat, D. (1989), 'Know-how, Innovation, and Regional Development', mimeo, Institut de Recherches Économiques et Régionales, Université de Neuchâtel.

Dosi, G. (1982), 'Technological Paradigms and Technological Trajectories', *Research Policy*, 11: 147–62.

Elsasser, H. (1982), 'Sectoral Shifts in Economic Structures: An Overview', in E. Brugger et al. (eds.), *The Transformation of Swiss Mountain Regions: Problems of Development between Self-Reliance and Dependence in an Economic and Ecological Perspective* (Berne: P. Haupt).

Glasmeier, A. (1991), 'Technological discontinuities and flexible production networks: The case of the world watch industry', *Research Policy*, 20: 469–85.

—— and Sugiura, N. (1991), 'Japan's manufacturing system: Small businesses, subcontracting, and regional complex formation', *International Journal of Urban and Regional Research*, 15/3: 395–414.

Hainard, F. (1988), *Savoir-Faire et culture technique dans l'Arc Jurassien* (Neuchâtel: UNESCO).

Harrison, B. (1993), 'The devolution of the Italian districts', in *Big Firms, Small Firms: Corporate Power in the Age of Flexibility* (New York: Basic Books).

Katzenstein, P. (1984), *Corporatism and Change: Austria, Switzerland and the Politics of Industry* (Ithaca, NY: Cornell University Press).

—— (1985), *Small States and World Markets: Industrial Policy in Europe* (Ithaca, NY: Cornell University Press).

Kelley, M., and Brooks, H. (1988), *The Case of Computerised Automation in US Manufacturing* (Cambridge, Mass.: Harvard University Press).

Knickerbocker, A. (1974), *Notes on the Watch Industries of Switzerland, Japan and the United States (Abridged)* (Boston, Mass.: Harvard Business School).

—— (1976), *Notes on the Watch Industries of Switzerland, Japan and the United States (Revised)* (Boston, Mass.: Harvard Business School).

Landes, D. (1979), 'Watchmaking: A case study of enterprise and change', *Business History Review*, 53/1: 1–38.

—— (1983), *Revolution in Time: Clocks and the Making of the Modern World* (Cambridge, Mass.: Belknap Press).

—— (1984), 'Time Runs Out for the Swiss', *Across the Board*, 21/1: 46–55.

Maillat, D. (1982) (ed.), *Technology: A Key Factor for Regional Development* (Saint-Saphorin: Georgi Publishing Company).

—— (1984), 'De-industrialization, tertiary-type activities, and redeployment', Paper presented at the 24th European Congress of the RSA, Milan (Dossiers: Université de Neuchâtel).

—— (1986), 'New technologies from the viewpoint of regional economics: The case of Neuchâtel', Paper presented at the International High-Tech Forum Seminar, Neuchâtel (Dossiers: Université de Neuchâtel).

—— (1988a), 'The case of the Franco-Swiss border from Geneva to Basle', *Transatlantic Colloquy on Cross-Border Relations: European and North American Perspectives* (Zurich: Schulthess Polygraphischer Verlag).

—— (1988b), 'Transfrontier regionalism: The Jura arc from Basle to Geneva', in I. D. Duchacek, D. Latouche, and G. Stevenson (eds.), *Perforated Sovereignties and International Relations* (New York: Greenwood Press).

—— (1989), 'Local dynamism, milieu and innovative enterprises', Paper presented at the Conference 'L'Innovation dans l'entreprise: Le Contexte spatial et culturel,' BALSTAHL, 2–3 Oct. (Neuchâtel: Université de Neuchâtel).

—— and Vasserot, J. (1988), 'Economic and territorial conditions for indigenous revival in Europe's industrial regions', in P. Aydalot and D. Keeble (eds.), *High Technology Industry and Innovative Environments: The European Experience* (London: Routledge).

Sabel, C. (1989), 'Flexible specialisation and the re-emergence of regional economies', in P. Hirst and J. Zeitlin (eds.), *Reversing Industrial Decline? Industrial Structure and Policy in Britain and her Competitors* (Oxford: Berg).

Sayer, A. (1989), 'Post Fordism in Question', *International Journal of Urban and Regional Research*, 13/4: 666–95.

Scott, A. (1988), *From Metropolis to Urban Form* (Los Angeles: University of California Press).

Storper, M., and Harrison, B. (1991), 'Flexibility, hierarchy and regional development: The changing structure of industrial production systems and their forms of governance in the 1990s', mimeo, Department of Planning, UCLA, Los Angeles.

—— and Walker, R. (1990), *The Capitalist Imperative* (Oxford: Basil Blackwell).

Union Bank of Switzerland (1986), *The Swiss Watchmaking Industry* (UBS Publications on Business, Banking and Monetary Topics, 100, UBS, Zurich).

—— (1987), *Economic Survey of Switzerland* (Zurich: UBS).

Vasserot, J. (1986a), 'L'Encouragement économique des régions en Suisse: L'Exemple de l'Arc Jurassien' (Institut de Recherches Économiques et Régionales, Université de Neuchâtel).

—— (1986b), 'Les Milieux innovateurs: Le Cas de l'Arc Jurassien en Suisse', mimeo, Institut de Recherches Économiques et Régionales, Université de Neuchâtel.

7

Regulating Labour:
The Social Regulation and Reproduction
of Local Labour-Markets

Jamie Peck

. . . the idea of a self-regulating market implied a stark utopia. Such an institution could not exist for any length of time without annihilating the human and natural substance of society; it would have physically destroyed man and transformed his surroundings into a wilderness (Polanyi 1944: 3).

In his analyses of the degrading and degenerative effects of the self-regulating market, Polanyi mounted an attack on those liberal doctrines which associated markets with freedom and regulation with the denial of freedom. Polanyi's critique of nineteenth-century capitalism demonstrated that—contrary to the nostrums of free-market ideology—the development of markets for 'fictitious' commodities such as labour was dependent upon the existence of a set of institutional forms which had the effect of countering, or containing, the self-destructive dynamic of the market system (1944: 68–76). This later led to the revelation that economic structures are, in different ways, 'socially embedded', that economic processes are 'instituted processes' (Polanyi 1957: 243–5). This institution-alization of economic processes, moreover, was seen to be the source of stability and unity in the political-economic system as a whole (ibid. 246–50). Understanding the way in which an economy was institutionally embedded, then, provided the key to understanding how that economy was able to reproduce itself.

It is perhaps not surprising that these insights should be rediscovered and re-evaluated at a time when liberal discourses have passed through a second ascendancy. In recent years Polanyi's ideas have become increasingly influential in institutionalist economics and in economic sociology (see Zukin and DiMaggio 1990; Mendell and Salee 1991; Granovetter and Swedberg 1992). Just as significantly, these ideas also have strong resonances in the growing body of work which is seeking to excavate the social, cultural, and political foundations of economic growth and restructuring (see Aglietta 1979; Offe 1985; Lash and Urry 1987; Boyer 1990*b*; Sayer and Walker 1992). Yet while a diverse group of writers is beginning to come to terms with the fact that 'institutions matter', the questions of why and how much they matter continue to raise controversy. The argument that certain types of institution are essential for economic growth is, of course, perilously close to functionalism: we can all enjoy local economic growth if we clone the 'successful' institutions of, say, the Third Italy. True, institutions have effects, but rarely, if ever, are these guaranteed.

An enduring quality of Polanyi's work is that a strong case is made for the necessity of institutional regulation which is at the same time an explicitly anti-functionalist one. Structural flaws in self-regulating markets are identified, which in turn pose 'regulatory problems', but these problems may be tackled in a variety of institutional ways. The 'need' for regulation, then, does not determine the form of the institutional response. This question of 'institutional indeterminacy' is an important one for contemporary debates around regulation theory, which has a tendency to assert, rather than demonstrate, the need for social regulation and is consequently justifiably exposed to accusations of functionalism (Tickell and Peck 1992). Whilst regulation-school approaches provide some challenging insights, the concept of social regulation is itself rather underdeveloped, abstract in definition, and macroeconomic in application (Marden 1992; Peck and Tickell 1992*b*).

This chapter is an initial attempt to bridge some of these gaps through an exploration of labour-market regulation. The point of departure is the question, 'why regulate labour?' Following this, a regulationist framework is deployed to examine—first in a rather abstract way—the relationship between social regulation and space. A framework for examining social regulation and uneven development is then proposed, within which the specific question of local labour-market regulation is examined. Here it is argued that labour-market regulation is associated with a variety of institutional forms and that these forms are geographically differentiated. These vary from locality to locality, region to region, even within highly centralized national regulatory regimes such as that of the United Kingdom.

The chapter will argue that labour-markets are regulated in different ways in different places. Local institutions—broadly defined—play a key role in labour-market reproduction. This is illustrated in an examination of different institutional responses to the generic 'regulatory problem' of skill formation. In the case of the labour-market, then, not only do institutions matter, but local institutions matter. The differing ways in which, and means through which, local labour-markets are socially regulated has real implications for local economic destinies. The issue of social regulation consequently needs to be treated as part and parcel of the problematic of uneven development.

WHY REGULATE LABOUR?

Regulation and markets, in effect, grew up together (Polanyi 1944: 68).

The idea of a self-regulating labour-market is a fiction. In essence, this is because labour itself is a 'fictitious' commodity: it is not produced for sale, it cannot be stored, it cannot be separated from its owner, it is 'only another name for human activity which goes with life itself' (ibid. 72). To treat labour as a commodity and the labour-market as a market is to isolate labour from the social relations in which it is embedded (Block 1990; Peck 1992a), to ignore the fact that both the production and reproduction of labour are intrinsically social processes. It follows that the labour-market cannot be self-regulating in the sense of an abstract commodity market, but must be socially regulated.

Jettisoning the conception of the labour-market as a self-regulating commodity market unfortunately also means abandoning simple descriptions of how labour-'markets' operate. The labour-market is an immensely complex social structure, an amalgam of social processes. In reality, there is no single 'logic' of the labour-market to set against the simple elegance of the economists' supply-and-demand schedules. The labour-market, for example, cannot be explained solely in terms of a 'production logic', in terms of the requirements of capitalist production, for this cannot account adequately for the ways in which work is divided up along race and gender lines. Labour-market segmentation theory is certainly the most advanced in developing a grasp of the richness and complexity of 'real world' labour-markets, forwarding a multicausal explanation of labour-market structure and dynamics. The generative structures of the labour-market are traced to three 'families' of social processes in contemporary segmentation theory: production imperatives and the associated design of jobs and structuring of labour demand which follow from these; processes of social reproduction and the structuring of the labour supply; and forces of regulation, with particular emphasis on the role of the state. Each of these generative structures exerts a particular influence upon the ways in which labour-markets are structured (for a discussion of this, see Peck 1989). In this sense, labour-markets are 'conjunctural' phenomena, the composite result of a variety of intersecting social processes.

Production-centred explanations have emphasized the role of techno-logy, industry structure, and labour-control imperatives in the structuring of the labour-market. In the early dualist theories of segmentation, the burden of explanation was placed upon technological requirements and divergent development in industry structures (Doeringer and Piore 1971). The 'radical school' took this a step further, situating the idea of dualism within a broader historical context and drawing attention to the role of labour segmentation as a control strategy (Reich *et al.* 1973; Gordon *et al.*

1982). Through 'divide and rule' strategies, capital was seen to be using segmentation as a means for overcoming the political contradictions of deskilling and workforce homogenization.

The response to this demand-side orientation came in the form of assertions of the importance of, first, collective action on the part of trade unions, and second, the wider sphere of social reproduction in the determination of labour-market structures (Rubery 1978; Picchio del Mercato 1981; Humphries and Rubery 1984). This 'reconstitution' of the supply side of the labour-market demonstrated that labour-market structures do not simply reflect the interests of capital, but—as arenas of struggle, compromise, and accommodation—also bear the imprint of supply-side influences. Modes of collective action in the workplace, the organization of domestic labour, and the role played by the family in socializing the rising generation, for example, all exert influences on labour-market structure. Importantly, these supply-side structures, while certainly strongly conditioned by the demand side, are relatively autonomous from the specific nature and requirements of the job structure (Humphries and Rubery 1984).

Because the mechanisms of social reproduction on the supply side of the labour-market have to be seen as relatively autonomous—related to, though not determined by, those of the demand side—it follows that the labour-market cannot be self-regulating. The system of social reproduction is affected by—but not governed by—the 'iron laws of supply and demand', with the result that the supply of labour to the market becomes, in effect, a 'regulatory problem'. What is it that determines that the appropriate quantities and qualities of labour-power are supplied to the market, when the processes governing the reproduction of labour are so much more complex than utility maximization? As Polanyi (1944) explained, the act of labour's 'coming into being' as active wage-labour is largely unrelated to calculations about its subsequent saleability on the market: labour, unlike genuine commodities, is not just 'produced for sale'. The reproduction of labour is a social and demographic process. The 'regulatory problem', then, is that of incorporating labour-power into the labour-market. This process of transforming '*dispossessed* labour-power into active wage-labour', Offe and Lenhardt have argued, 'was not and is not possible without state policies' (1984: 93).

Offe and Lenhardt go on to identify three aspects of this regulatory problem: the preparedness of those 'outside' the labour-market to offer their labour-power on the market; the circumstances under which non-participation in the labour-market can be permitted; and the balance between the waged and unwaged sections of the population. Processes of work socialization in the family and in the education system have a key role to play in the formation of a working class motivated to participate in the labour-market. These processes are, however, complex, and it is crude to say the least to argue that they are all articulated directly by the state (compare Bowles and Gintis 1976; Willis 1977). Nevertheless, the state certainly performs a 'policing' function in this area, as the recent policy responses of British governments in the areas of youth training and vocational education clearly illustrate (see Rees and Rees 1991; Peck 1992*b*).

The state is more deeply implicated in the regulation of conditions surrounding non-participation in the labour-market. Decisions as to whether or not to engage in wage-labour cannot be left to the individual worker, but must be closely regulated so as to ensure the effective minimization (given the prevailing labour-demand conditions) of non-participation in the labour-market (Offe and Hinrichs 1985). This is achieved through the delimitation of those social groups with a legitimate 'alternative role' outside the labour-market (usually those involved in social reproduction and those who due to age or incapacity are unable to make a significant contribution to production) and through the degrading and punitive conditions under which such non-participation is permitted (necessary to ensure that non-participation is not perceived as an 'attractive' alternative to wage-labour). The boundary between participation and non-participation is not, of course, static, but is constantly shifting along with changing conditions in the economy. It is typical in periods of high labour demand for the state to pursue strategies for the release of 'non-participants' on to the labour-market (for example, through the extension of child-care provision), while in periods of low labour demand the capacity of social catchment areas outside the labour-market may have to be extended, through such means as the deployment of make-work schemes and the withdrawal of state provision from the care sector (see Labour Studies Group 1985; Wilkinson 1988). Moreover, with structural change in the labour-market, such as has been seen in many parts of Europe since the 1970s, it may become necessary to exclude certain social groups from the labour-market on a more or less permanent basis (Hinrichs *et al.* 1988).

Some might dismiss this account of the imperatives for state regulation of the labour-market as an example of state-functionalism. To a certain extent, this charge is warranted, despite the fact that its proponents recognize that the precise form of the state's response is not determined by the regulatory need which is being satisfied. This is in fact held to be the product of class struggles and compromises (de Brunhoff 1978; Offe 1985). While this insistence removes this work one step from straightforward functionalism, its approach nevertheless remains resolutely statist. Here, regulatory needs invariably call for state responses of some kind. Regulation theory, in contrast, has a much richer conception of social regulation, encompassing both state and non-state processes of regulation (see Boyer 1990*b*). The means through which social regulation is achieved are not functionally preordained, but may be realized (or partially realized) through a plethora of institutional forms. Thus, although the state continues to play a central role, regulatory needs may be 'satisfied' through either state or non-state means. This regulationist conception of social regulation is now examined in greater detail.

REGULATION THEORY, SOCIAL REGULATION, AND SPACE

Social Regulation in Regulation Theory

Regulation theory holds that capitalist development occurs through successive 'regimes of accumulation', punctuated by periods of crisis and restructuring. The concept of a regime of accumulation refers to a relatively stable phase of economic development and comprises two elements: an accumulation system (the dominant mode of economic growth and distribution, with its characteristic labour processes, patterns of consumption, configuration of inter-firm relationships, etc.) and a mode of social regulation (an ensemble of state forms, habits, and social norms which underwrite the system of accumulation, containing its inherent crisis tendencies over at least the medium term).

Regulation theory has been popularized through analyses of Fordism–Keynesianism, the dominant regime of accumulation in most of the advanced industrial countries in the first three decades of the post-war period, which coupled an intensive accumulation system (based upon mass production and mass consumption) with a monopolistic mode of social regulation (based upon the Keynesian welfare state and institutionalized forms of income distribution). Despite the intense speculation which has surrounded the crisis of Fordism–Keynesianism and the possible emergence of a successor, concepts of 'post-Fordism' remain only partially developed from a regulationist standpoint. Post-Fordism is an asymmetrically developed concept because, for all the work on the emergent form of flexible accumulation, comparatively little has been said about flexible regulation (Peck and Tickell 1992*a*; Tickell and Peck 1992).

Regulation theory too is open to attacks of functionalism. Some have argued that the cluster of social norms, political practices, and state forms that make up the mode of social regulation (MSR) seem to be ushered in (or established by the state) to meet the regulatory needs of the accumulation system and therefore that these institutions themselves are functionally determined (Teague 1990; Clark 1993). This charge is denied by regulationists (see Lipietz 1988), who insist that modes of social regulation are formed through (indeterminate) political and social struggles, while the establishment of a stable coupling between an accumulation system and an MSR is a 'chance discovery' (Lipietz 1987; Jessop 1990).

The formation of MSRs—and therefore of sustainable regimes of accumulation—is contingent upon the vagaries of national politics: 'struggle and institutionalized compromises tend to arise within the framework of individual nations; hence the methodological priority [in regulation theory] given to . . . *primacy of internal causes*' (Lipietz 1987: 21–2).

In addition to the methodological priority afforded to the national social formation, regulation theorists also tend to emphasize particular aspects of the MSR, notably the regulation of wage and labour relations. This focus on wage and labour relations arises not from any inherent priority placed on these issues in the architecture of the theory itself, but because historical analysis has deemed these key points of tension in the process of regulatory restructuring and crisis. As Boyer (1988: 10) explains:

[The regulationist approach] is sufficiently broad for us to be able to anticipate *a priori* close linkages between the form of wage/labour relations and the method of regulation . . . [To a considerable extent] economic crisis and change in wage/labour relations determine one another.

The burden of explanation in regulationist accounts has consequently been placed on the regulation of struggles around the labour process and income distribution. As Brenner and Glick (1991) have asserted in their critique of regulationist work, this implies that, in practice, regulationist explanations are narrower and more partial than the theory's pretensions.

To a certain extent, such criticisms are warranted. As a research project, regulation theory requires further development if it is to fully grasp the complexities of capitalist growth and crisis. One of the key issues here is the regulationists' treatment of space and spatial scale. Regulation theory is constructed in essence from national building blocks, this being the geographical scale at which regulatory struggles, crises, and experimental 'solutions' are seen to be rooted. Regimes of accumulation, however, are also shown to become dominant at the international scale (see e.g. Lipietz's (1987) analysis of global Fordism). Regulation theory can be viewed, first and foremost, as a theory of the restructuring of national economies (along with their supporting institutional frameworks), and secondly and less explicitly, of how these articulate with the global economy. This is somewhat problematic, though, for concepts such as Fordism, which are often used interchangeably at the national and global scales. It would seem that, while global Fordism is viewed as the 'generic' version of the regime, there are 'beneath' this a series of 'national mutations'—from 'racial Fordism' in South Africa to 'blocked Fordism' in the United Kingdom, from 'peripheral Fordism' in Mexico to 'flex-Fordism' in Germany. In effect, these represent nationally specific—and hence geographically unique—'sub-couplings' of systems of accumulation and modes of social regulation (Peck and Tickell 1992*a*).

Despite the scope for indeterminate political struggles around the definition of national MSRs, only those MSRs which are ultimately compatible with the requirements of the accumulation process will subsequently stabilize under a regime of accumulation—effectively an a posteriori functionalism (Hirst and Zeitlin 1991). Taking this a step further, some critics of regulation theory have argued that all institutional struggles occur within the parameters established by global accumulation:

the given international distribution of productive power will have a central role in determining what institutions are even viable within national economies at a given historical conjuncture, as well as what will be their effect on capital accumulation, since, unless they are shielded in some way, these institutions must directly respond to international competition. . . . [The] economic viability and effects of . . . modes of regulation will heavily depend on the stage of development of the world economy . . . No doubt, *defined within that context*, institutions have proved, and will continue to prove, extremely important in affecting regional or national paths of growth of the productive forces . . . variations in institutional forms across nations and regions will, in other words, have a major part in determining the hierarchies of productivity and competitiveness among regions and nations (Brenner and Glick 1991: 111–12).

While it is certainly true that regulatory struggles and institutional change are 'bounded' processes, Brenner and Glick overstate their case. They hold up the world economy as the source of inevitable—and apparently unregulated—economic forces. For these writers, institutions are the object of on-the-ground 'local struggles', while the real direction derives from the ethereal imperatives of global economic restructuring. The world economy is thus 'naturalized' (see Block 1990), presented as if it were somehow analytically prior to 'domestic' economies and in some way independent of politics.

While this analytical subordination of the process of social regulation may simply be a case of turning regulation theory's a posteriori functionalism back on itself, it is nevertheless mistaken on three grounds. First, it is wrong to imply that the global economy is unregulated, as the international sphere has always been deeply regulated, initially largely through the economic imperialism of Pax Britannica and Pax Americana, and latterly and increasingly through the formation of supranational state structures (see Polanyi 1944; Marchak 1991; Ruccio *et al.* 1991). Second, it is from a very narrowly functionalist–instrumentalist position that Brenner and Glick (1991: 111) argue that institutions are 'constructed so as to emulate or surpass institutions in place elsewhere in promoting competitiveness in production'. This is hardly an adequate basis on which to interpret, for example, local-government restructuring, health policy, or trade-union politics. Third, it is somewhat puzzling to assert that, while institutional processes 'have a major part in determining the hierarchies of productivity and competitiveness among . . . nations' (ibid. 112), they apparently have a negligible effect in shaping the global economic order itself.

In reality, accumulation and regulation interpenetrate at all spatial scales. Moreover, because the social structures of accumulation and regulation are relatively autonomous, yet bound together in a necessary relation, the causal liabilities with which they are endowed will be realized in different ways in different times and places, depending upon contingent circumstances. One might expect, then, that the nature of the regulation–accumulation relationship is qualitatively different at each geographical level. This proposition will be examined with reference to the question of subnational uneven development, with a view to situating the role of regional economies—and local labour-markets in particular—within a regulationist framework.

Social Regulation and Uneven Development

While a large body of work has explored the tendencies for uneven development which are implicit in the accumulation process (see Harvey 1982; Smith 1984; Massey and Allen 1988; Storper and Walker 1989), the question of the relationship between uneven development and social regulation has gone largely unexamined. This is a question which regulation theory itself has yet to confront (Peck and Tickell 1992*a*). I want to suggest that the relationship between social regulation and uneven development is articulated at three levels: first, the uneven development of the economy itself must be regulated, in the sense that the nationally dominant MSR must be capable of containing the tensions and contradictions following from the uneven subnational distribution of wealth, employment, and access to resources; second, components of social regulation, particularly state policies, produce uneven spatial effects, as an intentional or accidental consequence of their design; and third, processes of social regulation will contingently result in uneven spatial effects because of the way in which they interact with historically prior uses of space (i.e. the pre-existing geographies of accumulation and regulation).

To begin with the first of these issues, uneven development must itself be regulated in the sense that a functional MSR must be capable of containing the geographical contradictions of the mode of growth. Each mode of growth, of course, tends to be associated with a particular form of uneven spatial development: regimes of accumulation tend to be associated with a distinctive spatial core and a distinctive spatial periphery, a geography visible at both the subnational and international scales (Webber 1982). Thus, the geographical core of the Fordist regime centred upon the heavily industrialized regions of the north-eastern United States and north-western Europe, while the periphery was constituted of subordinate regions both within the Fordist core countries themselves and within the newly

industrializing countries. Within the core countries, the contradictions of
Fordist uneven development were contained by the monopolistic MSR of
Keynesian-welfarism, with its interventionist regional policies and extensive welfare system. The remorseless peripheralization of production under
Fordism, driven by the search for cheap labour costs, subsequently played
a role in destabilizing production in the core and ultimately undermining
the regime itself (Lipietz 1986).

In the period of political and economic experimentation which has
characterized 'after Fordism', new growth models have emerged—each
with their own modes of uneven development—though none has yet
demonstrated the durability necessary to stabilize under a new regime.
Thatcherism, for example, proved incapable of sustaining the late 1980s
growth pattern in the United Kingdom. Though some saw in Thatcherism
the seeds of a sustainable MSR (see Jessop *et al.* 1988; Jacques and Hall
1989), this politico-economic project has not been able to contain its
internal contradictions. Significant amongst these are the geographical
contradictions of Thatcherism, popularized in the 'north–south divide'
debates of the late 1980s, but materially manifest in the collapse of the
south-east economy in the early 1990s recession (Peck and Tickell 1992*a*,
1992*b*).

A second aspect of the relationship between social regulation and
uneven development concerns the production, notably by state policies, of
spatially uneven effects. In terms of state action, policy initiatives produce
uneven spatial effects for one of two reasons: first, and most straightforwardly, some policies are explicitly designed as spatially targeted
interventions (for example, regional policies, enterprise zones, urban
development corporations); second, other policies produce uneven spatial
effects as an inevitable consequence of their operation, though nominally
they may appear to be 'aspatial' in design (for example, the spatial
concentration of production which results from EC competition policy, the
uneven geographical impact of higher-band tax cuts and mortgage relief,
the differential spatial effects of small-firm policies). These implicit and
explicit spatial tendencies are non-trivial because they produce qualitatively
different effects in different places. Differential policy impacts are not
simply matters of quantitative degree (region A receives a larger share of
national-defence expenditure than region B), but also—and significantly—
of qualitative effect (because of the role played by defence spending in
region A, it has evolved a particular industrial structure—robust during
cyclical recessions, but vulnerable to structural decline—along with
particular forms of production politics).

It follows that, because state policies produce uneven spatial effects,
their functions may also be geographically differentiated. For example,
youth training programmes schemes were shown in the 1980s to have
performed quite different functions in the buoyant economies of the south

of England and in the depressed regions of the north. In the context of expanding labour-markets with a high rate of vacancy generation, these policies performed the primary function of subsidizing employment, of underwriting the costs of recruitment, induction, and initial wages, while in depressed areas their primary function was to contain unemployment in a comparatively cost-effective manner through the provision of 'alternatives' to employment (Peck 1990). In Offe and Berger's (1985) terms, the predominant function of these ostensibly 'aspatial' policies, then, was of labour-market 'inclusion' in the buoyant labour-markets of the south (where the principal effects were in terms of the redefinition of recruitment and work norms) and labour-market 'exclusion' in the depressed north (where the programmes were used to redefine the system of welfare support).

Geographies of state policy are therefore important, not just as deviations from some notional 'national average', but as an integral aspect of the architecture, functions, and effects of the policies themselves. Similarly, though perhaps less evidently, spatially uneven functions and effects also characterize non-state social regulation. The much-vaunted 'enterprise culture', for example, had material and ideological resonances in the south-east of England which it could never have had in the deindustrializing north, with its quite different economic structure and political norms (see Keat and Abercrombie 1991; Shields 1991). The imposition of the ideology of southern growth—with its emphasis on individualism, entrepreneurship, and self-help—was one of the ways in which the north was subordinated, and its problems redefined, in the Thatcher era.

The third and final set of relationships between social regulation and uneven development are those 'interaction effects' which result from the interplay between regulatory process and historically prior uses of space. At a somewhat lower level of abstraction than the above-mentioned relationships, these interaction effects result from the way in which regulatory processes are realized as concrete events in particular local contexts. For example, while a particular policy programme may carry with it, by virtue of its nature, a set of implicit spatial tendencies, whether and how these tendencies are realized as empirical outcomes will vary from place to place, depending upon the interaction with pre-existing social structures. The case of training policy can again be used to illustrate the point.

The local consequences of 'national' training policies will depend on the overall architecture of the policy package, but also upon the particular configuration of institutional structures and interests at the local level. The local effects of the British youth training scheme, YT, for example, vary partly as a result of the ways in which the institutional machinery of the programme interacts with pre-established institutional structures (such as those developed under the earlier policy regimes of the Industry Training Boards and Youth Opportunities Programme) and with the particular configuration of labour–capital–state relationships at the local level (Peck and Haughton 1991). Similarly, the local character and policy priorities of the recently established Training and Enterprise Councils (TECs) have been shown to be related to spatially variable factors such as the institutionalization of business politics, the role of individual business leaders and key companies, the history of relationships between the public and private sector, the structure of the local economy, and the prior nature of the local training infrastructure (Peck 1991, 1992*b*). Despite the powerful centralizing tendencies in the TEC system (deriving from the contractual and financial framework within which individual local TECs operate), these factors have been shown to have significant material effects upon TEC structures, policy 'styles', and modes of programme development (Peck and Emmerich 1991; Emmerich and Peck 1992).

To sum up, there is a complex and multifaceted relationship between social regulation and uneven development, such that it is ill-advised to talk of some set of ubiquitous functions and effects of social regulation. Rather, these functions and effects will inevitably vary from place to place. The geographical anatomy of regulatory systems results not just from their internal tendencies and casual liabilities, but also from the complex and indeterminate ways in which these are reconciled with other, simultaneously unfolding social processes, and with contingent interactions with prior uses and space. So, while at an abstract level we can assert that the accumulation system and the MSR must couple together harmoniously for a stable regime of accumulation to be formed, this coupling will itself take a spatially variable form, the precise nature of which cannot be determined a priori. In this sense, any such coupling is a 'chance discovery' (Lipietz 1987).

LABOUR-MARKET FLEXIBILITY, SOCIAL REGULATION, AND SPACE

The Question of Labour-Market Flexibility

While there is widespread agreement that labour-market flexibility (variously measured) is increasing, no such consensus exists over either its causes or its consequences (for a discussion of these issues, see Hudson 1989; Peck 1992*a*). Through the 1980s the term 'labour-market flexibility' was appropriated by a number of different (and sometimes opposing) ideological discourses and was often applied rather casually in a variety of politico-economic contexts. Often, and particularly in policy circles, labour-market flexibility was held up as the desired 'end state', when in fact the label more accurately refers to an ongoing process of labour-market restructuring. This said, several quite different restructuring strategies have been encompassed within the generic of labour-market flexibility. At the crudest level, these can be arranged between the three poles of competitive, structured, and post-agrarian flexibilities. *Competitive flexibilities* have been associated with labour-market restructuring in the United Kingdom and involve the selective deregulation of the labour-market (specifically, the erosion of social protection for the unemployed, disadvantaged, and low-paid), the establishment of more individualized employment relations (as collective representation and bargaining structures have broken down), and the exposure of the workforce to sharpened competitive pressures. Secondly, *structured flexibilities* tend to be associated with more 'organized' labour-markets, such as those of Germany which combined a high wage structure with well-developed forms of social protection, collectivized political structures, and a high degree of qualitative (i.e. cross-skill) flexibility in the workplace. Finally, *post-agrarian flexibilities* characterize the development path of many southern European labour-markets, which are able to combine flexible forms of wage-labour (notably, part-time and seasonal work) with farm and family work systems, employment conditions typically being highly exploitative in each sphere.

In far more subtle ways than this, a host of commentators have pointed to variability in the nature of labour-market flexibility (Lash and Urry 1987; Mahon 1987; Leborgne and Lipietz 1988, 1990; Amin 1989; Brunhes 1989; Rojot 1989; Amin and Robins 1990; Boyer 1990a; Hadjimichalis and Papamichos 1990; Michon 1990; Sengenberger and Pyke 1990). What these writers in their different ways are underlining here is the importance of social structures and regulatory mechanisms in not just setting the context, but shaping the character of local economic development processes. The nature of labour-market flexibility is seen to be a product of the local social and regulatory milieu in which it is embedded. Thus, while labour-market flexibility can be associated with the development of progressive work norms in one context, it can mean poverty, exploitation, and insecurity in another (see Hudson 1989).

Such approaches, tending as they do to emphasize the local specificities of labour-market flexibility, stand in sharp contrast to those which identify a common underlying dynamic to these local development processes (e.g. Piore and Sabel 1984; Scott 1988a, 1988c; Storper 1990; Storper and Scott 1990). These writers link the formation of flexible labour-market structures to a posited transition towards flexible production systems. This transition, which is often characterized—prematurely from a regulationist point of view (Tickell and Peck 1992)—in terms of the establishment of a regime of flexible accumulation, is held to be the bearer of a distinctive spatial logic: the reagglomeration of production into networks of specialized industrial districts. In Scott's *New Industrial Spaces* (1988c), this spatial logic is rendered explicit. Deploying a transactions costs approach, Scott argues that the need for enhanced flexibility in production systems has led to a process of dynamic vertical disintegration, in which firms seek to become more responsive and flexible by externalizing many of the functions previously performed within the organization. This, in turn, implies a deepening of the social division of labour as firms become enmeshed in dense networks of inter-firm linkages and labour-market relations. It is this high degree of interdependence in flexible production systems which induces a process of spatial agglomeration: on the one hand, firms seeking to minimize the costs of inter-firm transactions will be induced to cluster together in space, while on the other hand, the need to establish new, more flexible, labour-market norms will also lead to agglomeration around particular local labour-markets (Scott 1988c: 27–31; cf. Henry 1992).

In this account, then, tendencies for agglomeration arise from the interdependence of specialized producers and from the functionality of particular labour-market structures. There are evidently strong echoes here of Marshall's classic account of industrial districts, emphasizing as it did the benefits of a 'constant market for skill' and the advantages

which people following the same skilled trade get from near neighbourhood to one another. The mysteries of the trade become no mysteries; but are as it were in the air, and children learn many of them unconsciously. . . . And subsidiary trades grow up in the neighbourhood, supplying it with implements and materials, organizing its traffic, and in many ways conducing to the economy of its material. . . . [These] industries, devoting themselves each to one small branch of the process of production, and working it for a great many of their neighbours, are able to keep in constant use machinery of the most highly specialized character; and to make it pay its expenses (1890: 332).

For writers such as Scott, the search for flexibility places a premium on these economies of agglomeration. With specific reference to labour-market processes, the need for flexible labour is seen to engender spatial agglomeration.

These labour-market agglomeration tendencies, Scott (1988b: 120–38) explains, follow from four interrelated characteristics of large labour-markets. First, labour turnover rates tend to rise along with the size of the local labour-market: employers of contingent labour tend to be attracted to large, volatile labour pools because in these locations workforce adjustments can readily be made through alterations to hiring and firing policies. Second, and from the supply side, the virtue of this volatility for contingent workers is that, although individual job tenures may be unstable, the constant turnover of vacancies means that unemployment durations are usually short. Third, the greater the size of a local labour-market, the lower is the marginal cost of search activities, for both firms and workers. Fourth, volatile labour-market conditions tend to become socially embedded, as workers are progressively socialized into particular work rhythms. Through such means, the agglomeration tendencies described by Scott effectively become self-perpetuating.

While these arguments are in many ways persuasive, they fall short of taking account of the broader—though critical—issues of the reproduction and regulation of flexible labour agglomerations (Peck 1992*a*). While Storper and Scott (1989: 25) note the coincidence of this posited shift towards flexible accumulation and the presence of governments 'committed to varying degrees to attempts to dismantle the apparatus of Keynesian welfare-statism', they stop short of explaining how flexible labour-market forms can be regulated or reproduced. This is a significant omission because flexible labour-markets are shot through with internal tensions and contradictions which, if they are not contained in some way, can place a brake on the development process. Witness, for example, the collapse of the fragile growth pattern established in the south-east of England, where poaching, wage inflation, skill shortages, and malfunctioning in the housing market contributed to the chronic overheating of the regional economy (Peck and Tickell 1992*b*). It is apparently a common error to mistake *strategies* for labour flexibility (some of which will be experiments doomed to failure) with reproducible flexible labour-market *structures*. Such 'readings off' from contemporary labour-market strategies to putative future labour-market structures can be misleading. It is consequently important to focus less on 'newness' *per se*, than on the key issues of the reproduction and regulation of labour-market structures.

Reproduction and Regulation in Flexible Labour-Markets

For flexible labour-markets to be demonstrably reproducible, they must be capable of containing their internal tensions and contradictions. These can be enumerated as follows (see also Peck 1992*a*). First, flexible labour-markets face potential problems in the reproduction of skills: as internal labour-markets are dismantled and labour externalized, so existing structures of skill formation are undermined. In situations where labour mobility is high and firm sizes generally small, one might expect under-investment in skill formation to follow, as firms—if they can carry the initial costs of training—are unlikely to invest in the training of workers who are subsequently likely to leave. Flexible labour-markets, though viable in the short term, are consequently prone to incipient skills crises (Peck and Haughton 1991). Second, high levels of external labour flexibility produce problems of labour control, as direct supervision must be replaced by 'arms-length' workforce control systems (Michon 1987). While strategies such as outworking, for example, can be deployed in order to circumvent shopfloor organization, they present other labour control problems and engender a degree of dependence upon external labour which effectively places limits on the division of labour. Third, the development of qualitative flexibility within the core workforce (through strategies such as multi-skilling and team-working) creates other potential labour-control problems, as core workers' bargaining strength within the labour process may be enhanced and as the 'tradability' of these skills on the external labour-market is increased. Fourth, the widening of social inequality which follows from highly dualized flexible labour-markets can lead to alienation and social conflict (Rubery *et al.* 1987; Teague 1990), such that the political reproduction of the system is compromised. Predictable social tensions emerge as the exploited periphery finds itself living cheek-by-jowl with the privileged core (see Saxenian 1983). Fifth, by virtue of their spatial proximity—and possibly also by virtue of shared labour-market and workplace experiences—flexible labour agglomerations may prove susceptible to (new forms of) collective labour organization (see Lane 1988; Mathews 1989).

Such internal contradictions point to the fragility of flexible labour-market growth and underscore the importance of developing appropriate systems of social regulation. The contradictions of flexible labour-markets pose 'regulatory dilemmas' (Selznick 1985; Marden 1992), medium-term 'solutions' for which must be found/happened upon if growth is to be sustained. Of course these solutions will tend to be only partial, temporary, and indeed are likely in themselves to stimulate further tensions. For example, a collectivized training system may be developed in order to overcome some of the problems of skill formation in flexible labour-markets, but once constituted this system is likely to become the object of social struggles around finance, training content, qualifications, and so forth (see Perry 1976).

For regulation theory, the existence of such regulatory 'needs' (albeit articulated in different ways in different contexts) is no more than the initial trigger for the search for a regulatory 'solution': the regulatory need does not determine the form of the institutional response. In this sense, the theory stops short of outright functionalism, but retains the idea that certain regulatory functions must be performed *in whatever way* in order for economic growth to be sustained (compare Peet 1989; Michon 1990; Teague 1990; Clark 1992; Marden 1992). What becomes important, then, is not so much that these generic regulatory needs exist, but how they are *responded to* in different temporal and spatial contexts. The contrasting ways in which the question of the social regulation of skill formation has been tackled under conditions of competitive, structured, and post-agrarian flexibility are illustrative of this.

The Case of Skill-Formation

The British case illustrates the limits of competitive flexibility strategies in the labour-market. The breakdown of growth in southern England in the early 1990s recession was in part a consequence of the problems of regulating the process of skill-formation under competitive flexibility.

Under the clarion call of 'deregulation', the British training system was substantially restructured during the 1980s with the result that state support was effectively withdrawn for all but the most basic, transferable-skills provision for the unemployed (Finegold and Soskice 1988). Employer-led TECs, launched in 1988 to develop locally tuned training responses, lack the financial or organizational resources to respond to anything but the most basic skill needs and still cater to a predominantly unemployed client group. Ironically, because the TECs inherited from the government the responsibility for unemployment programmes, they are at their weakest in buoyant areas of the country were unemployment levels have been historically low (Peck 1992*b*).

The absence of any legislative or organizational framework for inter-mediate and high-skill training in Britain has left firms dependent upon internal structures of skill formation (where feasible) and the haphazard process of poaching. The dismantlement of the statutory training frame-work in Britain, and its replacement with a market-orientated and voluntarist system, has destabilized inter-firm relations in the labour-market. 'Deregulation', in this sense, has meant the 'law of the jungle': the self-destructive dynamic of the market was released. Accelerating skilled labour turnover, deepening skill shortages, and spiralling wage inflation have been the predictable consequences, particularly in sectors unable to sustain internal labour-markets, such as the construction industry and parts of the financial services. These difficulties had by the late 1980s reached endemic proportions in the booming local economies along the M4 and M11 corridors (Murray 1989). Symptomatic of the wider problem of over-exploitation of the region's social, economic, and physical infrastructures during this period, this training crisis was part and parcel of the wider crisis of regulation which was eventually to contribute to the suffocation of the south's growth pattern (Leadbetter 1991; Peck and Tickell 1992*b*).

Certainly in as far as skill-formation is concerned, the deregulationist approach of the British government was destined for failure. As Streeck (1989) has explained, such strategies fail to recognize the character of skills as 'collective goods': firms pursuing their individual 'rational' self-interest will tend to under-invest in skills because they cannot be assured of realizing the full value of training investments (particularly when they are surrounded by other firms who have made the same calculation and concluded that poaching is the solution). The answer to this dilemma of skill-formation is, according to Streeck, not to return wholeheartedly towards state provision of training (which is often inadequately relevant to workplace needs), but to instigate a state-regulated, neocorporatist approach along the lines of the German 'dual system':

While market motives and processes play a significant and recognized part in [this] system, they are controlled by, and embedded in, what are essentially collective agreements between monopolistic employers associations and trade unions exercising, under a state licence, delegated public responsibility which enables them to impose effectively binding obligations upon their memberships. . . . [The growing needs of] firms for collective, non-appropriable production factors, like a rich supply of high and broad functional and extrafunctional skills, opens up political arenas where corporatist self-government of social groups may be a superior mode of regulation compared to both state intervention and the free market (Streeck 1989: 103).

The dual system, which is highly formalized and combines elements of workplace-based training with off-the-job provision in training schools, is widely recognized to have made a significant positive contribution to the productivity of German industry (Steedman and Wagner 1987; Prais and Wagner 1988). The system is based upon extended apprenticeships (usually of three and a half years' duration) and has been substantially reformed through the 1980s so as to deliver more broadly based skills and greater capacities for skilled worker autonomy.

The German training system has a strong regional component: while the broad framework for workplace training is a national one, training schools are under the control of individual *Länder*, which have also tended to develop distinctive systems of complementary institutions and policy programmes. Baden Württemberg, for example, enhances its basic regional training infrastructure with complementary programmes for the support of innovation and technology transfer (Cooke and Morgan 1990*a*, 1990*b*). These regional differences are, Rees (1990: 72) has argued, now increasing as strains on the dual system have led to 'increasing disparities between experiences at the level of the firm, the region and the individual'. These tensions in the German training system are essentially twofold. First, acceleration in the rate of technological change and the concomitant shortening of product life cycles—particularly in dynamic regional economies such as Baden Württemberg—has reduced the effective 'shelf life' of technical skills (Cooke and Morgan 1990*a*). This, in turn, has produced an increased demand for continuing training, which varies both qualitatively and quantitatively and between regions, sectors, and firms because it is overwhelmingly workplace-based and beyond the scope of state regulation (Rees 1990; Mahnkopf 1992). Second, this need for continuing training, while accommodated by most large firms, poses organizational problems for the SME and craft sectors. These sectors have begun to explore new institutional means of satisfying their continuing skill needs, such as training consortia, partnership arrangements, and subcontracting to specialist training agencies (Hoch 1988). As a consequence, further unevenness is emerging in the training system.

While Germany retains one of the most robust systems for initial and basic training, there is increasing evidence that the cumbersome process of institutional reform is lagging—perhaps even disrupting—the process of technical and labour-process change, particularly in the more dynamic sectors of the economy (Rees 1990). The process of redesigning and implementing regulations for training in the metal trades, for example, took virtually ten years to complete. This is leading to a situation in which some firms and some sectors are becoming partially disfranchised from the dual training system, which remains primarily oriented to the needs of large employers. Such formal, institutional approaches to skill-formation, under conditions of structured flexibility, are by no means universal panaceas, but create their own tensions and contradictions.

The situation is different again under the post-agrarian flexibility characteristic of areas such as Emilia-Romagna. In classic accounts of the development of small-film networks in this region, much is made of the role of local institutions and the virtues of spatial clustering in the development and exploitation of a rich skills base (see Amin and Robins 1990). As these and other writers have pointed out, however, the initial, Marshallian-style growth of flexible production in Emilia-Romagna was predicated upon the prior existence of traditional agrarian social structures (Amin 1989; Camagni and Capello 1988; Hadjimichalis and Papamichos 1990). Thus, many of the skills deployed in these paradigmatic flexible production systems were initially appropriated from the domestic or agricultural spheres. This 'dependent' mode of skill-formation was later partially superseded by collaborative training institutions, where SMEs cooperated with one another (and often also with local state and business associations) in the establishment of more formalized structures (Perulli 1990). Such responses to the 'collective good problem' posed by skill-formation were, however, only partially successful. Collective goods such as work skills are not simply 'in the air' in a Marshallian sense, but have to be socially produced and reproduced. The difficulties in Italy arose because, Trigilia argues, most responses were local and partial, and not effectively coordinated at the regional scale:

the role assumed by forms of flexible productive organization poses a *regulative problem of a regional nature*. These forms require, in fact, innovations of an organizational and technological character, managerial and entrepreneurial training, the training, reskilling and mobility of the labour force. . . . [This requires] more so than in the past, an 'intermediate government' of economic development; a regional dimension which is certainly not capable of managing directly, but rather of stimulating and co-ordinating relations between different public and private actors, and also between areas with different characteristics. . . . Technological innovation, training and labour mobility . . . are supplied with difficulty through bureaucratic structures which lack the consent, information and collaboration of the interested parties. Only authoritative representative organizations, at a level wider than the local, could 'internalize' the benefits derived from efficient regional intervention and would therefore have a greater incentive to contribute to the realization of such interventions (Trigilia 1991: 313–15).

Because the regional tier in Italy is comparatively weak, no such formal institutionalization has occurred. The mode of growth exhibited by Emilia-Romagna is also a fragile one (though for completely different reasons to the British case), because it is predicated upon a unique—and now changing—social structure and because the institutions for socio-economic reproduction are only partially formed.

As Hadjimichalis and Papamichos (1990) have demonstrated, the social structures which fostered flexible production are now beginning to break down: there is a growing reluctance to accept irregular and 'semi-legal' employment, the cultural values of agrarian society are being eclipsed by the lures of modern consumerism, traditional family structures are decaying as women begin to challenge their subordinate roles in both the domestic and wage-labour economies, and the old systems of political patronage are being eroded by economic modernization. Ironically, then, the very 'successes' of the regional economy—in particular its progressive integration into the global economy and its inculcation with global competitive imperatives—have begun to undermine the bases of its initial competitiveness. Emilia-Romagna now faces the challenge of overcoming these 'social blockages' and re-creating new forms of local social solidarity and cooperation within the context of international competition (Cooke and Morgan 1991). This, to be certain, is an exceptionally delicate operation, which may prove effectively impossible to realize through the application of comparatively blunt policy instruments.

To sum up, responses to the regulatory dilemma of skill-formation have been quite different in Britain, Germany, and Italy, underlining the differences in the nature of labour-market flexibility found in these different situations. The point here is not just that *regulation matters*, that it has important, material effects on the nature of the accumulation process, but that *regional regulation matters*. Systems of social regulation are unevenly developed across space, and not only in situations where regulatory norms seem clearly to have emerged from the 'bottom-up' (such as in Emilia-Romagna), but this assertion also seems to hold in those 'top-down', state-articulated regulatory regimes of Britain and to a lesser extent Germany. Even centrally designed and nationally imposed policy programmes such as the labour-market interventions of Conservative governments in Britain have very different effects in different places and play a part in the formation of distinctive regional 'regulatory milieux'. These, in turn, influence the form, structure, and dynamism of local labour-markets and shape the course of regional economic development.

While the 'reregulation' of the British training system, for example, may have played a role in the polarization of the skill and wage structure in the expanding economies of the south of England (by underwriting secondary-sector recruitment, induction, and employment and by exposing the higher echelons of the primary sector to external labour-market forces, labour poaching, and wage inflation), a different set of forces was unleashed in the high unemployment areas of the north (where the principal impacts occurred in terms of the containment of unemployment and the unpicking of the welfare safety-net). Similar arguments about the regionally specific character of regulatory systems and 'regulatory effects' can be developed for Italy (where the unique regional social structures of Emilia-Romagna conditioned the nature of the flexible production and labour-market structures which emerged there) and for Germany (where regionally articulated policy synergies and the spatially variable interaction between technical–labour process imperatives and the rigidities of the dual training system play a part in the formation of distinctive regional economic trajectories).

Mobile or Rooted Regulation?

To what extent can regulatory structures be said to be 'portable' from place to place? At a superficial level, forms of state regulation (state institutions, policy systems) may appear to be potentially more portable than those non-state forms of regulation, such as the particular function of the family and of agrarian traditions in the operation of Emilian labour-markets, which seem so self-evidently 'rooted in place'. But state forms too are less mobile than they may seem, given that, while it may be possible to clone particular institutional structures or imitate particular policy programmes, it is quite a different matter to reproduce the effects of state regulation, because of the indeterminate way in which regulatory effects are realized in different spatial contexts. If the German dual training system were to be transplanted into Britain, for example, its precise institutional form and its economic effects would vary greatly from one area of the country to another, depending upon its interactions with spatially variable phenomena such as the nature of the local 'class bargain' between labour and capital, the prior nature of local states and local training infrastructures, the character of public–private-sector relations, the structure of local economies, and so on.

While some regulatory institutions may, in broad terms, be duplicable, their effects most certainly are not. This is because the process by which such effects are generated is dependent upon contingent local conditions. We can therefore think in terms of 'rooted regulation': even though most regions face a similar set of regulatory dilemmas, the nature of the institutional response to these will tend to be regionally idiosyncratic. Different regions may attempt different regulatory 'experiments', while national (and for that matter, international) experiments will take different forms in different places. This implies that geographical diversity itself becomes a non-trivial issue. Indeed, an appreciation of the role of spatial unevenness becomes a precondition for understanding the meaning and significance of regulatory experiments themselves. As Salais and Storper (1992: 189) have observed,

new forms of 'strategic alliance' in a variety of producer and consumer goods industries . . . , the 'corporatist' industrial systems found in metal working and machinery building in Bavaria and Baden Württemberg . . . and the 'contractual' production networks typical of American high technology industry . . . [are all] means of trying to resolve tensions introduced by rapid changes in products, the increasing complexity of the division of labour, and existing worlds of production and their [regulatory] conventions. In the process, new conventions are being developed and with them, perhaps, new worlds of production. A significant implication of this . . . is that there are important local and national (and not only sectoral) variations in such experiments. This adds yet another dimension to the industrial diversity which we expect from the existence of different kinds of products: the locally- and regionally-specific development of conventions (which may not be easily diffusable to other places).

Experimentation with regulatory systems and projects—conventions in Salais and Storper's approach—is likely, then, to be an ongoing process, though one which is particularly pronounced during the interstices of regimes of accumulation. Regulatory 'solutions' discovered in particular local situations, however, may not be generalizable to other contexts, even within the same nation state.

It is probable, moreover, that labour regulation will be particularly 'locally embedded', because labour itself—due to the social nature of its production and reproduction—is the most place-bound of the 'factors of production'. The production and reproduction of labour-power is dependent upon the supportive effects of certain key social institutions (family structures, schools, recreational organizations, and the like) and, as a consequence, requires a substantial degree of stability. The result of this is a 'fabric of distinctive, lasting "communities" and "cultures" woven into the landscape of labour' (Storper and Walker 1983: 7). Following this logic, the social institutions which underpin the regulation of labour-markets are themselves distinctively local. The labour-market is, indeed, one of the most socially—and in this sense also locally—embedded of economic systems.

SOME CONCLUSIONS

The argument presented here can be summarized as follows: labour-markets are not self-regulating; they consequently pose a series of 'regulatory dilemmas', such as the problem of skill-formation; the existence of a regulatory need does not determine the institutional form of the regulatory response; these institutional forms are geographically variable and subject to political struggles; in turn, their spatial unevenness contributes to the geographical variability in labour-market structures themselves; social regulation in general and the social regulation of the labour-market in particular are therefore 'rooted in place' and have material effects on the course of economic development. More broadly, it can be argued that accumulation and regulation will 'couple' together in different ways in different places, with differing degrees of functionality and with differing politico-economic consequences (Peck and Tickell 1992*a*). These different interactions, as Teague (1990: 41) has argued, may have the important effect of 'causing [national and regional] economies to develop along divergent paths'.

Regulatory systems are not portable structures, achieving similar results wherever they are deployed, but are in fact deeply rooted in local social structures. This assertion holds just as much for 'top-down' state institutions and policies—the effects of which vary from place to place—as it does for what are often perceived as the 'bottom-up' elements of modes of social regulation—custom, habits, behavioural norms, and the like. The geographies of regulation which are consequently brought into being subsequently exert non-trivial influences upon the course of accumulation in general and upon the nature of labour-markets in particular. Thus, it can be concluded that one of the root causes of the variable nature of labour-market flexibility follows from the fact that these labour-market structures operate within a variety of national, regional, and local regulatory milieux. Although one can deduce a priori that flexible labour-markets will be vulnerable to a series of generic regulatory dilemmas, the institutional responses to these will vary from one place to another, as subsequently will the diverse labour-market norms which evolve under the broad umbrella of 'flexibility'. In short, the social and spatial context in general and the regulatory milieu in particular will have a major influence on the 'type' of labour-market flexibility which emerges in a region. This is one of the reasons, then, why flexible labour-markets are associated with poverty in one place and prosperity in another, conflict in one place and cooperation in another, development in one place and decline in another.

The question of how, when, and where 'virtuous' relationships between accumulation and social regulation will be established remains, of course, an open one. The key issue here though has to be that of reproducibility: it is not enough just to accomplish economic growth, it has to be sustained. Whilst it is an evidently straightforward matter to distinguish between sustainable and unsustainable forms of growth retrospectively, it is inevitably much more difficult in contemporary research. But at least by focusing on questions of reproduction and regulation, and on how different regulatory dilemmas are being tackled institutionally, it might be possible to distinguish between some of the more durable and some of the more fragile forms of growth. This, in turn, should inform the process of separating out 'flash in the pan' new industrial spaces from those capable of sustaining medium-term growth.

From the perspective of the politics of regional economic development, there may yet be some way to go before a new generic mode of social regulation becomes hegemonic. As the south of England's burst of unsustainable development in the late 1980s has demonstrated, neoliberal approaches seem incapable of reproducing flexible growth. Perhaps more progressive 'structured' approaches to regulation are entering an ascendancy (Dunford 1990; Peck and Tickell 1992*a*). What is becoming clear from these experiences, though, is that there are unlikely to be any easy solutions. While local success stories continue to provide hope for local economic policy-makers (see Cooke and Imrie 1989; Luria 1990; Murray 1991), the evidence here is that regulatory solutions must be home-grown, not imported. For regions leached by years of neoliberalist deregulation, though, the concern has to be that the social subsoil may now prove to be barren.

The productivity of labor . . . like that of the soil . . . can build up over time, provided proper care is taken. The effect is to make each [local] labor market even more unique, because the long processes of sociopolitical development within an urban region can build up unique mixes of [labour] qualities. The plundering of those qualities through de-skilling, overworking, bad labor relations, unemployment, and so forth can, however, lead . . . to the rapid depletion of a prime productive force. . . . The problem, of course, is that the coercive laws of competition force individual capitalists into strategies of plundering, even when that undermines their own class interest. Whether or not such a result comes to pass depends upon the internal conditions of labor demand and supply, the possibilities of replenishing labor reserves through migration either of labor power or capital, and the capacity of capitalists to put a floor under their own competition by agreeing to some kind of regulation of the labor market (Harvey 1989: 133–4).

NOTE

The arguments put forward here have benefited enormously from discussions with colleagues on the European Science Foundation's RURE programme. An earlier version of the chapter was presented to the conference of the International Working Party on Labour-Market Segmentation at the University of Cambridge, July 1992. I would like to thank members of the Working Party, plus Nick Henry and Adam Tickell, for their helpful comments. Needless to say, they are absolved from responsibility for the chapter's contents.

REFERENCES

Aglietta, M. (1979), *A Theory of Capitalist Regulation: The US Experience* (London: New Left Books).

Amin, A. (1989), 'Flexible specialisation and small firms in Italy: Myths and realities', *Antipode*, 21: 13–34.

—— and Robins, K. (1990), 'The re-emergence of regional economies? The mythical geography of flexible accumulation', *Environment and Planning D: Society and Space*, 8: 7–34.

Block, F. (1990), *Postindustrial Possibilities: A Critique of Economic Discourse* (Berkeley, Calif.: University of California Press).

Bowles, S., and Gintis, H. (1976), *Schooling in Capitalist America* (London: Routledge and Kegan Paul).

Boyer, R. (1988), 'Wage/labour relations, growth and crisis: A hidden dialectic', in R. Boyer (ed.), *The Search for Labour Market Flexibility: The European Economies in Transition* (Oxford: Clarendon Press).

—— (1990a), 'The impact of the single market on labour and employment: A discussion of macro-economic approaches in the light of research in labour economics', *Labour and Society*, 15: 109–42.

—— (1990b), *The Regulation School: A Critical Introduction* (New York: Columbia University Press).

Brenner, R., and Glick, M. (1991), 'The regulation approach: Theory and history', *New Left Review*, 188: 45–119.

Brunhes, B. (1989), 'Labour flexibility in enterprises: A comparison of firms in four European countries', in *Labour Market Flexibility: Trends in Enterprise* (Paris: OECD).

Camagni, R., and Capello, R. (1988), 'Italian success stories of local economic development: Theoretical conditions and practical experiences', mimeo, Instituto di Economia Politica, Universita Luigi Bocconi, Milan.

Clark, G. L. (1992), ' "Real" regulation: The administrative state', *Environment and Planning A*, 24: 615–27.

—— (1993), *Pensions and Corporate Restructuring in American Industry* (Baltimore: Johns Hopkins University Press).

Cooke, P., and Imrie, R. (1989), 'Little victories: Local economic development in European regions', *Entrepreneurship and Regional Development*, 1: 313–27.

—— and Morgan, K. (1990a), *Industry, Training and Technology Transfer: The Baden Württemberg System in Perspective* (Regional Industrial Research Report, 6; Regional Industrial Research, University of Cardiff, Cardiff).

—— —— (1990b), *Learning through Networking: Regional Innovation and the Lessons from Baden Württemberg* (Regional Industrial Research Report, 5; Regional Industrial Research, University of Cardiff, Cardiff).

—— —— (1991), *Industrial and Institutional Innovation in Emilia-Romagna* (Regional Industrial Research Report, 7; Regional Industrial Research, University of Cardiff, Cardiff).

de Brunhoff, S. (1978), *The State, Capital and Economic Policy* (London: Pluto Press).

172   *Regulating Labour*

Doeringer, P., and Piore, M. J. (1971), *Internal Labor Markets and Manpower Analysis* (Lexington, Ky.: D. C. Heath).

Dunford, M. (1990), 'Theories of regulation', *Environment and Planning D: Society and Space*, 8: 297–322.

Emmerich, M., and Peck, J. A. (1992), *Reforming the TECs: Towards a New Training Strategy* (Centre for Local Economic Strategies: Manchester).

Finegold, D., and Soskice, D. (1988), 'The failure of training in Britain: Analysis and prescription', *Oxford Review of Economic Policy*, 4: 21–53.

Gordon, D. M., Edwards, R., and Reich, M. (1982), *Segmented Work, Divided Workers: The Historical Transformation of Labor in the United States* (Cambridg: Cambridge University Press).

Granovetter, M., and Swedberg, R. (1992) (eds.), *The Sociology of Economic Life* (Boulder, Colo.: Westview Press).

Hadjimichalis, C., and Papamichos, N. (1990), ' "Local" development in southern Europe: Towards a new mythology', *Antipode*, 22: 181–210.

Harvey, D. (1982), *The Limits to Capital* (Oxford: Basil Blackwell).

—— (1989), *The Urban Experience* (Oxford: Basil Blackwell).

Henry, N. (1992), 'The new industrial spaces: Locational logic of a new production era?', *International Journal of Urban and Regional Research*, 16: 375–96.

Hinrichs, K., Offe, C., and Weisenthal, H. (1988), 'Time, money and welfare-state capitalism', in J. Keane (ed.), *Civil Society and the State: New European Perspectives* (London: Verso).

Hirst, P., and Zeitlin, J. (1991), 'Flexible specialisation versus post-Fordism: Theory, evidence and policy implications', *Economy and Society*, 20: 1–56.

Hoch, H.-D. (1988), *The New Industrial Metalworking Occupations* (Berlin: Bundesinstitut für Berufsbildung).

Hudson, R. (1989), 'Labour-market changes and new forms of work in old industrial regions: Maybe flexibility for some but not flexible accumulation', *Environment and Planning D: Society and Space*, 7: 5–30.

Humphries, J., and Rubery, J. (1984), 'The reconstitution of the supply side of the labour market: The relative autonomy of social reproduction', *Cambridge Journal of Economics*, 8: 331–46.

Jacques, M., and Hall, S. (1989) (eds.), *New Times* (London: Lawrence and Wishart).

Jessop, B. (1990), 'Regulation theories in retrospect and prospect', *Economy and Society*, 19: 153–216.

—— Bonnett, K., Bromley, S., and Ling, T. (1988), *Thatcherism: A Tale of Two Nations* (Cambridge: Polity).

Keat, R., and Abercrombie, N. (1991) (eds.), *Enterprise Culture* (London: Routledge).

Labour Studies Group (1985), 'Economic, social and political factors in the operation of the labour market', in B. Roberts *et al.* (eds.), *New Approaches to Economic Life: Restructuring, Unemployment and the Social Division of Labour* (Manchester: Manchester University Press).

Lane, T. (1988), 'The unions: Caught on the ebb tide', in D. Massey and J. Allen (eds.), *Uneven Re-development: Cities and Regions in Transition* (London: Hodder and Stoughton).

Lash, S., and Urry, J. (1987), *The End of Organised Capitalism* (Cambridge: Polity).

Leadbetter, C. (1991), 'Britain's days of judgement', *Marxism Today* (June), 14–19.

Leborgne, D., and Lipietz, A. (1988), 'New technologies, new modes of regulation: Some spatial implications', *Environment and Planning D: Society and Space*, 6: 262–80.

—— —— (1990), 'How to avoid a two-tier Europe', *Labour and Society*, 15: 177–99.

Lipietz, A. (1986), 'New tendencies in the international division of labour: Regimes of accumulation and modes of social regulation', in A. J. Scott and M. Storper (eds.), *Production, Work, Territory: The Geographical Anatomy of Industrial Capitalism* (Boston: Allen and Unwin).

—— (1987), *Mirages and Miracles: The Crises of Global Fordism* (London: New Left Books).

—— (1988), 'Reflections on a tale: The Marxist foundations of the concepts of regulation and accumulation', *Studies in Political Economy*, 26: 7–36.

Luria, D. (1990), 'Automation, markets, and scale: Can flexible niching modernize US manufacturing?', *International Review of Applied Economics*, 4: 27–65.

Mahnkopf, B. (1992), 'The "skill-oriented" strategies of German trade unions: Their impact on efficiency and equality objectives', *British Journal of Industrial Relations*, 30: 61–81.

Mahon, R. (1987), 'From Fordism to?: New technology, labour markets and unions', *Economic and Industrial Democracy*, 8: 5–60.

Marchak, M. P. (1991), *The Integrated Circus: The New Right and the Restructuring of Global Markets* (Montreal: McGill-Queen's University Press).

Marden, P. (1992), ' "Real" regulation reconsidered', *Environment and Planning A*, 24: 751–67.

Marshall, A. (1890), *Principles of Economics* (London: Macmillan).

Massey, D., and Allen, J. (1988) (eds.), *Uneven Re-development* (London: Hodder and Stoughton).

Mathews, J. (1989), *Age of Democracy: The Politics of Post-Fordism* (Melbourne: Oxford University Press).

Mendell, M., and Salee, D. (1991) (eds.), *The Legacy of Karl Polanyi: Market State and Society at the End of the Twentieth Century* (Basingstoke: Macmillan).

Michon, F. (1987), 'Segmentation, employment structures and productive structure', in R. Tarling (ed.), *Flexibility in Labour Markets* (London: Academic Press).

—— (1990), 'The "European Social Community", a common model and its national variations? Segmentation effects, societal effects', *Labour and Society*, 15: 215–36.

Murray, R. (1989), *Crowding Out: Boom and Crisis in the South East* (Stevenage: South East Economic Development Strategy).

—— (1991), *Local Space: Europe and the New Regionalism* (Manchester Centre for Local Economic Strategies and South East Economic Development Strategy).

Offe, C. (1985), *Disorganized Capitalism: Contemporary Transformations of Work and Politics* (Cambridge: Polity).

—— and Berger, J. (1985), 'The future of the labour market', in Offe (1985).

174 *Regulating Labour*

—— and Hinrichs, K. (1985), 'The political economy of the labour market', in Offe (1985).

—— and Lenhardt, G. (1984), 'Social policy and the theory of the state', in C. Offe (ed.), *Contradictions of the Welfare State* (London: Hutchinson).

Peck, J. A. (1989), 'Labour market segmentation theory', *Labour and Industry*, 2: 119–44.

—— (1990), 'The Youth Training Scheme: Regional policy in reverse?', *Policy and Politics*, 18: 135–43.

—— (1991), 'The politics of training in Britain: Contradictions of the TEC initiative', *Capital and Class*, 44: 23–34.

—— (1992a), 'Labor and agglomeration: Labor control and flexibility in local labor markets', *Economic Geography*, 68: 325–47.

—— (1992b), 'TECs and the local politics of training', *Political Geography*, 11: 335–54.

—— and Emmerich, M. (1991), *Challenging the TECs* (Manchester: Centre for Local Economic Strategies).

—— and Haughton, G. F. (1991), 'Youth training and the local reconstruction of skill: Evidence from the engineering industry of North West England, 1981–1988', *Environment and Planning A*, 23: 813–32.

—— and Tickell, A. (1992a), 'Local modes of social regulation? Regulation theory, Thatcherism and uneven development', *Geoforum*, 23: 347–64.

—— —— (1992b), 'Time, space, flexibility: Uneven development in regulation theory' (Working Paper 92/18; School of Geography, Leeds University, Leeds).

Peet, R. (1989), 'Conceptual problems in neo-Marxist industrial geography: A critique of themes from Scott and Storper's *Production, Work Territory*', *Antipode*, 21: 35–50.

Perry, P. J. C. (1976), *The Evolution of British Manpower Policy—from the Statute of Artificers of 1563 to the Industrial Training Act of 1964* (London: BACIE).

Perulli, P. (1990), 'Industrial flexibility and small firm districts: The Italian case', *Economic and Industrial Democracy*, 11: 337–53.

Picchio del Mercato, A. (1981), 'Social reproduction and the basic structure of labour markets', in P. Wilkinson (ed.), *The Dynamics of Labour Market Segmentation* (London: Academic Press).

Piore, M. J., and Sabel, C. (1984), *The Second Industrial Divide: Possibilities for Prosperity* (New York: Basic Books).

Polanyi, K. (1944), *The Great Transformation: The Political and Economic Origins of Our Time* (Boston: Beacon Press).

—— (1957), 'The economy as instituted process', in id., C. M. Arensberg, and H. W. Pearson (eds.), *Trade and Market in the Early Empires* (Glencoe: Free Press).

Prais, S. J., and Wagner, K. (1988), 'Productivity and management: The training of foremen in Britain and Germany', *National Institute Economic Review*, 123: 34–47.

Rees, G. (1990), *Vocational Education and Training Systems: The challenge of the 1990s* (School of Social and Administrative Studies, University of Cardiff, Cardiff).

—— and Rees, T. (1991), 'Educating for the "enterprise economy": A critical review', in P. Brown and H. Lauder (eds.), *Education for Economic Survival: From Fordism to post-Fordism?* (London: Routledge).

Reich, M., Gordon, D. M., and Edwards, R. C. (1973), 'A theory of labor market segmentation', *American Economic Review*, 63: 359–65.

Rojot, J. (1989), 'National experiences in labour market flexibility', *Labour Market Flexibility: Trends in Enterprises* (Paris: OECD).

Rubery, J. (1978), 'Structured labour markets, worker organisation and low pay', *Cambridge Journal of Economics*, 2: 17–36.

—— Tarling, R., and Wilkinson, F. (1987), 'Flexibility, marketing and the organisation of production', *Labour and Society*, 12: 131–51.

Ruccio, D., Resnick, S., and Wolf, R. (1991), 'Class beyond the nation state', *Capital and Class*, 43: 25–42.

Salais, R., and Storper, M. (1992), 'The four "worlds" of contemporary industry', *Cambridge Journal of Economics*, 16: 169–93.

Saxenian, A. (1983), 'The urban contradictions of Silicon Valley: Regional growth and the restructuring of the semiconductor industry', *International Journal of Urban and Regional Research*, 7: 237–62.

Sayer, R., and Walker, R. (1992), *The New Social Economy* (Oxford: Basil Blackwell).

Scott, A. J. (1988*a*), 'Flexible production systems and regional development: The rise of new industrial spaces in North America and western Europe', *International Journal of Urban and Regional Research*, 12: 171–86.

—— (1988*b*), *Metropolis: From the Division of Labor to Urban Form* (Berkeley, Calif.: University of California Press).

—— (1988*c*), *New Industrial Spaces: Flexible Production Organization and Regional Development in North America and Western Europe* (London: Pion).

Selznick, P. (1985), 'Focusing organisational research on regulation', in R. Noll (ed.), *Regulation and the Social Sciences* (Berkeley, Calif.: University of California Press).

Sengenberger, W., and Pyke, F. (1990), 'Small firm industrial districts and local economic development: Research and policy issues', *Labour and Society*, 16: 1–24.

Shields, R. (1991), *Places on the Margin: Alternative Geographies of Modernity* (London: Routledge).

Smith, N. (1984), *Uneven Development* (Oxford: Basil Blackwell).

Steedman, H., and Wagner, K. (1987), 'A second look at productivity, machinery and skills in Britain and Germany', *National Institute Economic Review*, 122: 84–96.

Storper, M. (1990), 'Responses to Amin and Robins: Michael Storper replies', in F. Pyke, G. Beccatini, and W. Sengenberger (eds.), *Industrial Districts and Inter-Firm Co-operation in Italy* (Geneva: International Labour Office).

—— and Scott, A. J. (1989), 'The geographical foundations and social regulation of flexible production complexes', in J. Wolch and M. Dear (eds.), *The Power of Geography: How Territory Shapes Social Life* (Boston: Unwin Hyman).

—— —— (1990), 'Work organisation and local labour markets in an era of flexible production', *International Labour Review*, 129: 573–91.

—— and Walker, R. (1983), 'The theory of labour and the theory of location', *International Journal of Urban and Regional Research*, 7: 1–41.

—— —— (1989), *The Capitalist Imperative* (Oxford: Basil Blackwell).

Streeck, W. (1989), 'Skills and the limits of neo-liberalism: The enterprise of the future as a place of learning', *Work, Employment and Society*, 3: 89–104.

Teague, P. (1990), 'The political economy of the regulation school and the flexible specialisation scenario', *Journal of Economic Studies*, 17: 32–54.

Tickell, A., and Peck, J. A. (1992), 'Accumulation, regulation and the geographies of post-Fordism: Missing links in regulationist research', *Progress in Human Geography*, 16: 190–218.

Trigilia, C. (1991), 'The paradox of the region: Economic regulation and the representation of interests', *Economy and Society*, 20: 306–27.

Webber, M. J. (1982), 'Agglomeration and the regional question', *Antipode*, 14: 1–11.

Wilkinson, F. (1988), 'Deregulation, structured labour markets and unemployment', in P. J. Pedersen and R. Lund (eds.), *Unemployment: Theory, Policy, Structure* (Berlin: Walter de Gruyter).

Willis, P. (1977), *Learning to Labour: How Working Class Kids Get Working Class Jobs* (Aldershot: Gower).

Zukin, S., and DiMaggio, P. (1990) (eds.), *Structures of Capital: The Social Organisation of the Economy* (Cambridge: Cambridge University Press).

8

The Disembedded Regional Economy: The Transformation of East German Industrial Complexes into Western Enclaves

Gernot Grabher

INTRODUCTION: THE 'HALF REVOLUTION' IN
EASTERN GERMANY

Years after the events of autumn 1989, it appears that only 'half a revolution' took place in eastern Germany. Although the 'revolutionary subjects' of 1989 triggered the implosion of the old tired-out system, they played hardly any role in the creation of the new. Moreover, the revolutionary developments almost completely lacked innovative, future-oriented ideas (Habermas 1990: 181). In this vacuum, instead of the development of new social visions, the immediate implementation of the blueprint of western German society and economy rose to the top of the agenda. This obvious preference for the successful western German model was clearly endorsed by an overwhelming majority of eastern Germans in the 1990 elections, for various reasons such as the hope for a quick improvement in living conditions, distrust of all eastern German élites, and fear of regressive developments in the Soviet Union. This decision reduced the transformation of eastern Germany to a mere *cloning* of the western German institutional framework. The new economic and social institutions of eastern Germany were to be set up simply as branches of the western German institutions—at a speed and with a vigour, however, that precluded any self-organization. In political terms, this rigorous cloning has led to a subject-less society (Häussermann 1992: 4), a representative democracy in an apathetic society. In economic terms, the vacuum resulting from blocked economic and social self-organization was abruptly and vigorously filled by foreign actors, above all western investors.

This chapter outlines the strategies of western investors and evaluates their impact on regional development in eastern Germany. This development is now largely determined by the western German investors: only 10.7% of investment and 9.1% of job commitments related to the privatization of the eastern German economy are of non-German origin. However, the dialectics of the unification process have been such that there are grounds to justify the inclusion of western German investors into

the category of 'foreign investors'. Unification has come to encourage
cultural and political separation and even generate—at the moment when
it was intended to be destroyed—a distinct East German (GDR) ethnicity.
Partly for this reason the chapter starts with a glance backward at the
organization of production in the old GDR.

THE BUREAUCRATIC FAÇADE OF THE GDR ECONOMY

Central Planning and the Formation of the Kombinate

In the western economies of the post-war period, the organization of
production mainly followed the model of industrial mass production. This
model, however, was also the leading paradigm for the East European
economies: efficiency through economies of scale. In the GDR, this
economic *Leitbild* implied a thorough reorganization of traditional
regional and sectoral production patterns based around small-scale craft
production. Before World War II, industrial mass production did not play
a central role in both the western and the eastern part of Germany. In the
leading industrial centres of eastern Germany—Saxony and Thuringia—
production was primarily organized in small to medium-sized firms.
Typically, these firms, such as the printing-machine builders, printers, and
publishers in Leipzig, were tightly knit together and formed locally
concentrated sectoral clusters. Another typical example was the Jena-
based production cluster consisting of small mechanical engineering firms,
glassworks, and research departments of the local university, which
together formed the nucleus of the later renowned Carl Zeiss Jena Optik.
Although these regional clusters cannot easily be compared with the
industrial districts of today, there remain certain parallels such as the tight
horizontal and vertical linkages between independent firms which allowed
for a high degree of cross-fertilization.

The history of industry in the GDR begins with a sweeping attempt to
radically transform the traditional, craft-based production pattern. The
three-level system of central economic planning through industry ministries
and the confederations of state-owned firms (*Vereinigungen Volkseigener
Betriebe*, VVB) sought to achieve higher levels of efficiency through
industrial concentration and specialization: within the GDR no single
product would be produced simultaneously by two different firms. This
first attempt to improve the efficiency of production at the cost of demand
flexibility, however, was only of limited success. The chronic shortages of
intermediate goods and the poor reliability of suppliers—the recurring
theme of the forty years of GDR industry—reflected the limitations of the
central planning authorities to enforce their aims. Since suppliers usually
were assigned to a different ministry from that for final producers,

economic planning and coordination did not follow the logic of the value chain, but fragmented the interconnections from the raw material to the final product (Voskamp and Wittke 1990: 15).

At the end of the 1960s a second thoroughgoing attempt to increase economies of scale on a national scale resulted in the creation of the *Kombinate* (industrial complexes), which provided a new institutional framework for the process of concentration and specialization within individual industries. The intermediate level of planning, composed of the confederations of state-owned firms (VVB) was dissolved and the newly created *Kombinate* based on product value-chains were assigned directly to corresponding ministries. In order to enhance coordination between and control of the various stages of production, the main suppliers and R&D capacities were integrated into the *Kombinate* according to the principle of 'reproductive self-containment' (*reproduktive Geschlossenheit*). In this context, at the beginning of the Honecker era in 1972, the majority of remaining craft-based private firms, which accounted for 13% of net industrial production, became also integrated into the *Kombinate* (Deppe and Hoss 1989: 38).

After a first wave of formation of the *Kombinate* at the end of the 1960s and the beginning of the 1970s, covering about one-third of total industrial employment, there followed a second, all-encompassing, wave at the end of the 1970s. In the era of Honecker, the giant corporation, economies of scale, and technological progress became the mutual guarantors of economic growth. In 1989 industry in the GDR consisted of 126 centrally coordinated *Kombinate* with twenty to forty plants and more than 20,000 employees each on average. In addition, plants which were coordinated at the level of the district, as was the case, for example, in the construction or food-processing industry, were integrated into 95 *Kombinate*, each with 2,000 employees on average (Institut für angewandte Wirtschaftsforschung 1990). The concentration of production within the highly specialized, primarily vertically integrated, *Kombinate* allowed for larger batch sizes and, thus, favoured a shift towards larger production units. For example, while in the GDR less than 1% of the industrial workforce was employed in plants with less than 100 employees, in the Federal Republic of Germany the share amounted to 22% (OECD 1991: 84). Production within these units was organized on a quasi-Taylorist basis (Deppe and Hoss 1989: 92). At one level, the chronic shortage of intermediate goods and spare parts called for permanent *ad hoc* interventions and a high degree of flexibility on the shop floor and, hence, did not allow for the application of strict Taylorist work organization. On the other hand, the macro-economic prerequisites for Taylorist work organization, that is a well-functioning social division of labour and market cooperation at the level of the economy, were only partially fulfilled: cooperation between and within the geographically dispersed *Kombinate* (the Pentacon

Kombinat, for example, consisted of sixty-five widely scattered plants) was seriously hampered by the desolate condition of the transport and communication infrastructure in the GDR (Schwarz 1991: 9).

The dissolution of the intermediate level of the confederations of state-owned firms (VVB) led to a concentration of strategic sector-wide functions (e.g. planning, financial targets, price-setting, and trading decisions) within the central administrations of the *Kombinate*. The consequences of this process of concentration were twofold. First, the administration of the *Kombinate* came to hold considerable power *vis-à-vis* the respective ministries. Second, integration into the *Kombinate* implied a concentration of all central management functions (training, R&D, sales, and purchasing) within headquarter plants (*Stammbetriebe*). Due to the strategic (and political) importance of these *Stammbetriebe*, financial and technical resources were barely allocated to other plants in a *Kombinat*. Although initially the formation of the *Kombinate* increased the productivity of GDR industry it had, at least from a contemporary perspective, two disastrous consequences for the regions.

First, although this model of economic development, based on autarkic large mass producers, led to an increasing level of industrialization in the lagging northern and eastern regions and, hence, reduced the traditional North–South divide, it also favoured the deepening of already existing and new regional monostructures. As a consequence, in no less than fifty-four districts (of a total of 189 districts) the leading industry accounted for 40–60% of total employment in the district (Maretzke and Möller 1992: 156). Second, as a result of the internalization of economic interactions, the notion of the region as a supply-base for firms no longer had any economic meaning. Beyond the utilization of the local labour force, the individual plants of the *Kombinate* had no economic relation with the region in which they were located. The pre-existing rich tissue of intra-regional, forward, and backward linkages was torn apart and superseded by inter-regional linkages within the *Kombinate*. As a consequence, the basis for regional multiplier effects was destroyed. Through the internalization of all economic interactions, from the supply of raw materials to the production of the final product, the role of the regions as a source of agglomeration economies was truncated, i.e. economies that arise from a diversified regional economic structure and which are essential for the long-term adaptability of regions. In other words, the rationalization of production within the *Kombinate* and across regional boundaries, as a thoroughgoing attempt to increase the efficency of production at the cost of demand flexibility, ended up destroying a flexibility once provided by localized production clusters.

Behind the Bureaucratic Façade: Reciprocity and Barter
within Informal Networks

Behind the façade of the centrally coordinated *Kombinate*, however, lay anything but the 'precision, promptness, clearness, continuity, discretion, uniformity, strict subordination, savings on frictions, material and personal costs' celebrated by Max Weber (1972: 561) in defence of bureaucracy. The GDR economy corresponded with textbook models of bureaucratic planned economies about as much as do western European economies with textbook models of market economies. In the GDR, as in the other central and eastern European countries, resources were by no means allocated exclusively by the central planning authorities. In addition, informal exchange networks within and between the *Kombinate* played a key role in not only compensating for the chronic shortages of raw material, spare parts, and equipment, but also dealing with the continual *ad hoc* interventions of various power groups, such as local party members and trade unionists.

Compared, however, to other central and eastern European countries, the relative importance of these informal networks in the early GDR economy remained limited since the private sector compensated for the weaknesses of the shortage economy. However, in the 1970s the private sector largely lost this compensatory role. The recentralization of the economy in the course of the second wave of the formation of the *Kombinate* reduced the share of the private sector in total employment by two-thirds, from roughly 15% to about 5% (Deppe and Hoss 1989: 38). This severe restriction of private economic activities, which, after the beginning of the 1970s were confined to the retail and the craft sector, served, however, to increase the importance of informal networks, especially in the industrial supplies sector. These networks provided a diffuse infrastructure for barter governed by the principle of reciprocity. Reciprocity is a more general pattern of exchange than the principle of equivalence which supposedly governs market transactions, since exchanges are not expected to balance in every single act but over the entire exchange relation (Grabher 1993: 8). Thus, if a member of such an informal network received spare parts or equipment from another member of the network, he was not obliged to return the service immediately. However, the receiver was expected to assist other members of the network in a similar situation. To be sure, the exchanges did not involve only the supply of raw materials or spare parts, but also payment through the offer of labour or accommodation in the *Kombinat*-owned holiday homes. Although the larger part of such exchange took place in the grey area of personal networks reinforced by mutual obligations, some *Kombinate* turned it into an auxiliary organizational device: they circulated special 'pendulum lists' (*Pendellisten*) among different production sites of the *Kombinate* indicating the resources and capacities that were idle and of potential use as a buffer inventory to cope with unforeseeable shortages.

DISEMBEDDING THE EAST GERMAN ECONOMY AND ITS SOCIAL INSTITUTIONS

The peculiarity of the GDR economy was not based so much on the mere *existence* of a discrepancy between the bureaucratic façade and the informal networks behind it, as in the systemic importance of the latter (Heidenreich 1992). This systemic importance of informal networks, however, was revealed only after their complete destruction. The surprise of West German politicians and experts at the sudden implosion of the GDR economy after the introduction of the Deutschmark reflected the ignorance of their central role in 'getting the job done'.

These networks fell victim to the decision taken after unification to decompose and privatize the *Kombinate* by the *Treuhand*, the privatization agency in eastern Germany. This approach led to an abrupt separation of the individual plants of the *Kombinate*; a radical down-scaling or shutdown of departments such as R&D whose future financial returns could not be calculated precisely, but were crucial for long-term adaptability; and the separation of social facilities such as child care, hospitals, holiday homes, or sports clubs once tied to the *Kombinate*. Roughly three-quarters (9,988) of eastern German firms were privatized in September 1992 and stripped of their role as a central institution of social integration, and it is becoming more and more clear that this strategy has resulted in not only a dramatic loss of training and R&D capacities (Grabher 1992: 222), but also in the dissolution of basic social institutions which could have formed a nucleus for developing the institutional fabric of a modern (local) civil society.

This strategy probably also paralysed the potential for developing a social infrastructure for new economic activities. The loss of a supportive tissue goes beyond simply the loss of networks of personal ties, but represents, above all, the demise of entrepreneurial skills and experience related to the development of *ad hoc* solutions within the informal networks. It is unclear whether in reality the informal ties and the 'chaos-qualification' (Marz 1992: 9) embedded in these networks would have encouraged and supported start-ups. However, it is becoming all too clear that the rupture of 'old ties' and the subsequent atomization of economic and social actors did not lead to the effects for which it has been justified, namely to unleash market forces. This was an expectation based on the asociological[1] assumptions of classical and neoclassical economics, invoking, not distant from the Hobbesian concept of 'state of nature' or Rawls's 'original position', an idealized state of affairs in which economic behaviour and institutions remain untouched by social structure and social relations (Granovetter 1985: 481). However, the atomization of social and economic relations in eastern Germany has in no sense unleashed market forces. Instead, it has blocked the generation of indigenous economic activity.

In this context, great hope has been placed on inward investment by large western corporations. They are not only expected to transfer capital and the most advanced technical and organizational know-how to eastern Germany, but also to reshape the sectoral and spatial structure of its tired regional economies. Furthermore, they are expected to play an active role in constructing *ex novo* a new network of institutions such as chambers of commerce and trade associations supporting the emerging market economy. These rather ambitious expectations are based on the assumption that access to the eastern German and the wider eastern European market would motivate western investors to establish large facilities in eastern Germany to supply these markets. However, the assumption as yet has failed to materialize in most industries. First, trade with eastern Europe has collapsed since January 1991, when foreign trade came to be based on convertible currency. Second, western investors have been caught in the implosion of the eastern German productive system. Both these setbacks have forced most of the western investors to revise considerably their initial strategies—with rather ambivalent consequences for the eastern German regions.

THE IMPACT OF WESTERN INVESTMENT ON EASTERN
GERMAN REGIONS: RE-EMBEDDING THE EASTERN GERMAN ECONOMY?

The Merits of Eastern German Stones and Potatoes: Locally Integrated Production Complexes

Compared to other industries of the eastern German economy, the construction and the food, drink, and tobacco industries have been rather successful in attracting investors and in consolidating local production networks. This has occurred for two reasons. On the one hand, location decisions in these industries are largely influenced by the need to minimize high transport costs and the need for prompt delivery. Second, the enormous level of private demand and the large public investment programmes related to the improvement of the transport infrastructure (24.0 b DM in 1992), the federal railways (10.0 b DM), and housing construction (5.0 b DM) has been decisive for the take-over plans of western investors (Deutsches Institut für Wirtschaftsforschung 1992: 144). Amongst the more important western investors in the construction industry of eastern Germany are the RMC Group (UK) and Lafarge Coppée (France), who plan to invest 470 m DM and 350 m DM respectively to modernize cement plants and establish networks of distribution outlets for ready-made concrete (Morgan 1992: 4). However, also the smaller investment projects of the Italian RIVA Group and Feralpi SpA in the steel industry have to be seen in the context of the immense demand for bridges, railroads, and plant construction.

Since these plants serve local markets they enjoy a relatively high degree of local autonomy. They are equipped with basic managerial functions and sales and purchases departments. However, for technological reasons, the share of highly qualified managerial and technical staff is rather low. But, for the same reason, both forward and backward linkages within the region are relatively strong: the weight–price ratio of the basic materials of the construction industry does not allow for long transport distances. Some of the investors have even acquired shares of firms who once supplied the construction plants with raw materials such as gypsum and gravel. This strategy, which is not too far from the organizational logic of the *Kombinat*, might secure the survival of suppliers whose prospects as *Treuhand* firms were rather precarious (Deutsches Institut für Wirtschaftsforschung 1992: 141). This also may explain why especially in the first stages of the unification process a rather large share of investments was dedicated to the sand and stone industry. Already by 1991 western German corporations alone had invested approximately 2 b DM in this basic goods industry (Institut für Wirtschaftsforschung 1992: 9).

A similar pattern of corporate integration and regional embeddedness characterizes investment projects in the food, drink, and tobacco industry, which at present plays a key role in eastern Germany. Building on investments of 2.2 b DM in 1991, the investments of western German corporations will probably amount to 3.5 b for 1992. The significance of this scale of investment can be derived from the fact that the ratio between investments in eastern Germany and investments in western Germany within this industry is considerably higher than the average of 1 : 6 for manufacturing industry as a whole. Indeed, the development of the industry has to be seen in close connection with the strategies of the major western retail chains. The breathtakingly quick and nearly complete take-over of the eastern German retail and distribution sector by the major western German corporations Metro Group, Spar AG, Tengelmann Group, and REWE AG led to an equally breathtaking collapse in the eastern German food, drink, and tobacco industry. In the second half of 1990 production in the east Berlin food, drink, and tobacco industry, for example, dropped by 71.3% (Institut für Wirtschaftsforschung 1991: 43). Since the western retail chains maintain close relations with their western suppliers, eastern German producers had no chance of getting on to the order lists of the retail stores. However, this also reflected the sudden stigmatization of eastern products by eastern German consumers who preferred, regardless of quality and price, western products. However, this new consumer zeal of the eastern Germans was shortlived. One contributory factor was that the escalation of unemployment turned consumer preferences into a political issue: 'buying east' became more and more a demonstration of the disillusion with the unfulfilled promises of western capitalism. Another factor was the rediscovery by the eastern Germans, after a short period of experimentation with western products, of their customary liking of eastern German products.

The larger western investors in the food, drink, and tobacco industry— Coca Cola (USA), which committed investments of 700 m DM, Unilever (UK, 100 m DM), Philip Morris (USA), 60 m DM), and EAC (Denmark, 40 m DM)—tried to adapt to the local market with a twin strategy. First, they met the demand for popular western products by acquiring additional production facilities in eastern Germany. In a few cases, these corporations decided to establish, with much fanfare, new greenfield Euro-plants, which are dedicated to supply the EC market with a few Europe-wide brands (*Handelsblatt*, 31 December 1991). Second, and partially responding to the limits to any Europeanization of brands or any change in the preferences of eastern German consumers, they maintained the production of traditional eastern German products. The most prominent victim of these limits was the largest western German cigarette-producer, Reemtsma, which failed spectacularly to penetrate the eastern German market with western brands. In contrast, Philip Morris successfully pursued a strategy of 'regionalization' (Philip Morris jargon) by relaunching the most popular eastern German cigarette f6 with minimal modifications to the design and the material of the cigarette box (*Handelsblatt*, 4 December 1991). Imitating this strategy of 'regionalization', Reemtsma finally decided to relaunch the eastern German brands Cabinet and Juno, leaving their unparalleled flavour untouched. In the food industry, too, popular eastern German brands are celebrating a spectacular comeback. A handful of eastern German products, such as Nordhäuser Korn, a rather strong grain gin, were even promoted as German market leaders by their western German parent firm. In the largest western retail chain Metro AG, for example, the share of eastern German products in the total turnover realized in eastern Germany amounts to 10%; however, this share as well as the number of 300 eastern German products that are offered in western German retail stores are expected to grow (*Handelsblatt*, 2 February 1992).

From a regional point of view, this twin strategy has had important implications. As in the construction industry, the need to monitor the local market calls for a minimum level of management autonomy and marketing capability within the branch plants. This implies the creation of a tier of middle-management and qualified white-collar positions within rural labour markets suffering extraordinarily high levels of unemployment and massive deskilling of the labour force. The western investors also benefit from cheap inputs, especially in agricultural products, for which transport costs as well as the need for rapid delivery favour locational proximity. Indeed, the relatively high local content of meat, grain, and vegetable production may prevent the complete collapse of the eastern German agriculture. However, even the intense regional backward linkages of the food industry into the agricultural sector cannot prevent the massive loss of production capacity and skills in the industrial sector.

Eastern Pioneers of Lean Production? Post-Fordist Production Complexes in the Automobile Industry

New organizational developments in the automobile industry also favour the formation of regional backward linkages. According to the announcements of the two largest investors in eastern Germany in the passenger-car industry, Volkswagen and Opel (GM), supplier relations will resemble those characteristic of the most advanced just-in-time delivery automotive companies such as Toyota.

As the largest single manufacturing investment in eastern Germany, Volkswagen plans to erect a completely new automobile plant in Mosel near Zwickau, where from 1994 on 6,800 workers will produce 1,200 cars daily (Golf, type 3). Together with nearby supplier firms, Volkswagen intends to create 35,000 jobs (*Handelsblatt*, 26 September 1991). To achieve this ambitious aim, Volkswagen plans to invest 4.6 b DM over the next five years with 1.3 b DM contributed by federal budget grants (Lungwitz and Kreissig 1992: 179). Similarly, GM has committed itself to investing approximately 1 b DM to establish a plant at Eisenach, with an annual production capacity of 150,000 passenger cars (*Handelsblatt*, 13 December 1990). Like the Spanish GM plant in Saragossa, the chassis of the Opel models Corsa and Astra will be assembled at Eisenbach while the engine and the gear unit will be supplied from other European GM plants.

Both investors based their entry on the foundation of a 'joint corporation' in which they hold 12.5% of the shares. However, although the *Treuhand* owns 87.5% of the shares, a syndicate contract assures the western investors of the management of the 'joint corporation'. In fact, 18.5% of all investments in the manufacturing industries are based on such 'joint corporations', accompanied by strategies which sharply contrast with the rhetoric which surrounds western investors as pioneering, risk-taking capitalists. Since land and property relations within the 'joint corporation' are also included in cost-sharing calculations, the *Treuhand* has had to cover the bulk of the costs of preparing the sites and premises of the new greenfield plants of the western investors. These costs have included above all the costs of making good of past ecological damages, selecting and qualifying a workforce for the new plants and of financing assembly in the transition stage between the close down of lines devoted to the production of GDR cars and the opening up of the new plants. In other words, the 'joint corporation' forms part of a strategy to create an economic and social *tabula rasa* upon which can be erected a plant with a hand-picked élite of well-trained and highly motivated workers utilizing the most advanced technology—at the cost of the *Treuhand*. Eventually the western investors will fully own the 'joint corporation', and thereby assume all costs and risks, but not before this costly transition stage has been completed.

Initially, the western investors did not plan to close down the production of the old GDR cars completely. However, plans to maintain a small output of the passenger cars Wartburg and Trabant and of the trucks W50 and L60 for the eastern European market were abandoned when the *Treuhand* refused to cover the differential between production costs and sales price. The subsidy for the production of the Wartburg alone would have amounted to approximately 100 million DM annually (*Frankfurter Rundschau* 1991). Initially, Volkswagen also intended to partially shift the assembly of its smallest car Polo from its Spanish plant in Pamplona to the eastern German plant. However, as it became more and more clear that the rise of eastern German wage levels would make the assembly of small cars in Spain more profitable again in a relatively short period, Volkswagen decided to develop its eastern German branch as a major assembly plant for its compact car. Faced with the collapse of the market for eastern products and the prospects of diminishing wage differentials between eastern and western Germany, Volkswagen and GM decided to proclaim their eastern German plants as prototypes for the production of auto-mobiles of the future (Heidenreich 1992: 350).

GM, in particular, largely inspired by the crusade-like, anti-Japan advertising campaigns of its American headquarters lays claim to the adoption of the latest management fetish of 'lean production'. Although the rhetoric varies from corporation to corporation (Mickler and Walker 1992: 30), all major western car-producers seem to adhere to key elements of the corporate philosophy of Toyota, considered to be a winning formula to be beaten by its own standards. These include the decentralization of competences and responsibilities; the introduction of market elements within the corporate hierarchy; the reduction of the level of in-house production and the generalization of just-in-time supplier networks; the integration of production, maintenance, and quality control; and other celebrated new dogmas of automobile production (Womack, Jones, and Ross 1990). For the western German investors, eastern Germany appears as an almost perfect location for implementing this new 'best practice'.

Most importantly, the (vague) hope of getting a job in one of the prestigious western German corporations has facilitated a vigorous demolition of pre-existing work standards and individual aspirations, notably those regarding job security, frequency of changes in work organization, work intensity, etc. Western managers, indeed, revel in the possibilities of experimentation opened up—in the words of a manager—by the 'salutary cultural shock' to which eastern Germans have been exposed. This 'salutary cultural shock' allows them to introduce forms of work organization, which in the context of the highly institutionalized and negotiated system of industrial relations in western Germany would be much more troublesome to implement. Viewed from the perspective of western investors, the beneficial economic impact of social 'cleansing', however, must not be hindered by administrative and infrastructural backwardness. Thus, in the medium term, there is an enormous effort to renew the transport and telecommunication infrastructure as well as administrative structures, which will transform the former GDR into one of the infrastructurally most advanced production sites in Europe. A high-quality infrastructure is a basic precondition for the smooth integration of the eastern German plants into the European production networks of Volkswagen and GM.

The regional impact of the strategies of the large western investors is likely to be ambivalent. First, the implementation of new organization and management practices copying Toyota implies decentralized managerial competence at the operational level. The GM engine plant, for example, will be managed as a profit centre. However, all these plants will also be integrated within wider European corporate networks, with headquarters as well as the main research and development facilities outside eastern Germany (Mickler and Walker 1992: 42). Second, a key element of the concept of 'lean production' is a reduction in the ratio of in-house production. VW, for example, intends to achieve a ratio of between 25% and 30% in its plant in Saxony as compared to a ratio of 43% in its western plant at Wolfsburg (*Handelsblatt*, 26 September 1991). In order to encourage the development of a competitive regional supplier infrastructure and in order to guarantee Volkswagen quality standards, the company organized 'supplier conferences' to bring together pairs of eastern and western German producers of the same component. These conferences resulted in approximately forty take-overs and forty licensing agreements which will serve to supply the plant assembling the Golf type 3.

This strategy has enabled Volkswagen to shift the costs of monitoring and upgrading potential eastern suppliers to its western suppliers. Most probably, the eastern branches of the western suppliers will be integrated as second-tier suppliers within the supply pyramid controlled by large western first-tier suppliers (Doleschal 1991: 35–63). In any event, the eastern branch plants of western suppliers will not be equipped with their own R&D facilities. At best, they will be provided with small engineering departments for customer-specific adaptations of their products and for the development of special tools (Lungwitz and Kreissig 1992: 182). In addition the logistic competence of the eastern branch plants will have to be improved to meet the high requirements of just-in-time delivery. Most probably, the large eastern plants of Volkswagen and GM with their surrounding regional supplier networks will resemble the 'transplants' of the Japanese automobile producers in the United States: tightly integrated regional production complexes with extraordinarily high levels of quality and technological flexibility, but whose destiny is exclusively dependent on the strategy of one single corporation.

Cathedrals in the Eastern German Deserts: Modernizing Fordism in Mass Production Enclaves

In contrast to the above future-oriented experiment with post-Fordist concepts of organization, the majority of the western investors have preferred to opt for technologically advanced versions of rather familiar Fordist concepts. This strategy underpins investment in the chemicals, electrical engineering, metalworking, textiles, and clothing industries. It aims at combining the benefits of modern mass-production technology with the use of cheap, narrowly qualified or unskilled labour, monitored by technologically and organizationally most advanced means of corporate control. This forward-into-the-past strategy appears more as an *ad hoc* reaction to the collapse of the eastern European markets than as the result of long-term considerations. Several of the plants that have been taken over by western investors were initially planned as bridgeheads to these markets.

A case in point is the take-over of the Falkensee plant of the *Kombinat Outer Wear Berlin* by Helsa, a western German textiles producer. Following the disappointing development of the eastern market which shattered the strategy to establish a bridgehead, the western investor integrated its eastern German plant closely into its European network of production plants. The former activity of producing outerwear was closed down completely, production facilities were renewed, and streamlined down to the mass production of a rather simple textile component for outerwear (shoulder pads, which somehow seem to capture the essence of the current cultural mood in eastern Germany). The production of these components for men's outerwear has remained in the western plants, while the eastern German plant exclusively produces components for women's wear, a market which is much more contingent on seasonal and fashion cycles. Because the plant is restricted to a small stage in the production cycle and draws all its input from the western plants (which also receive the pads), all managerial functions have been run down. After the transition, overseen largely by western managers, only a foreman remains, in charge of ensuring that orders from the western headquarters, transmitted daily by Datex-p exchange, are met.

A large part of these Fordist attempts to utilize capacity in eastern Germany for cheap mass production or as a completely dependent source of supplies is based on wage agreements, an incentive which represents the most popular form of western integration into the manufacturing base (Institut der Deutschen Wirtschaft 1991: 6). The cost advantages of the eastern German plants do not only result from lower wages. (At present, for example, the tariffs in the chemical industry in eastern Germany amount to 55% of the western level. However, the real wages are considerably below this due to different Social Security regulations (Bispinck 1991: 22).) Of decisive importance, however, are the cost differentials resulting from the further fragmentation of the production process. At one level, this refers to the benefits resulting from the application of the technologically most advanced developments of the Babbage principle, that is, benefits from deskilling and related wage-reductions following from the further fragmentation of the production process. At another level, this fragmentation also allows some investors to escape from the highly regulated institutional environment of employer associations. The integrated production plants of the French Rhône-Poulenc group in western Germany, for example, are members of the employer association of the chemical industry. However, fragmentation of the production process has allowed Rhône-Poulenc to locate a few simple processing operations, now ascribed to the employer association of the textile industry, in eastern Germany. This guarantees, independently of the wage differentials between eastern and western Germany, permanent cost savings of approximately 15–20%, due to different agreed wage-tariff levels in the chemicals and the textile industries.

Clearly, the regional impact of the eastern German plants that are integrated into a wider corporate hierarchy on the basis of fulfilling a restricted number of low-order tasks, will be disappointing. They will probably remain 'cathedrals in the desert'. Since they are vertically integrated into the production chain of their parent corporation, they create limited regional supply opportunities and thus reduce the potential for multiplier effects within the region. As the survival of small, newly established firms frequently depends on the regional market, these 'cathedrals in the desert' constitute an impediment to the differentiation of a region's sectoral structure. In addition, regional linkages, particularly backward linkages, are the most important channel through which technological and organizational change is transmitted between firms. But, in the Cottbus plant of ABB, for example, the ratio of inputs supplied by western plants of the corporation amounts to approximately 80%. It is only in the area of construction and maintenance services that the eastern German electrical engineering plant of ABB draws inputs from local suppliers. Without doubt, the eastern plants of ABB, Rhône-Poulenc, or Helsa employ the most advanced production technologies and implement stringent quality control systems, but they fail to offer the transfer or demonstration effects which local firms need in order to raise their technological status and improve their organizational structure (Dicken 1990).

The truncated functional status of the 'cathedrals in the desert' has also important consequences for the stimulation of entrepreneurship. The lack of decision-making functions, especially those related to technical, scientific, and management tasks, is also an absence of the 'seedbed' that produces future entrepreneurs. It hinders the development of a sizeable regional middle class and the 'culture' associated with this stratum (Massey 1983: 66), and in turn reinforces the difficulty of attracting upper-echelon technical and managerial staff to other firms in a region—a vicious circle. In addition, the type of work organization within the 'cathedrals in the desert' seems not to be conducive for the creation of a regional *Mittelstand.* Applying the Babbage principle neither allows for a type of work organization that stimulates the social competences which are the main ingredients of future entrepreneurship nor for a further development of the chaos qualification that was acquired in the informal networks of the old production system.

TOWARDS A CAPITALISM WITHOUT CAPITALISTS? THE TRUNCATED INDUSTRIALIZATION OF EASTERN GERMAN REGIONS

The eastern German map of western industrial investment resembles a patchwork that is shaped by three core elements. The first element consists of loosely knit localized production complexes in the construction as well as the food, drink, and tobacco industries, which at present draw the major investors in eastern Germany. Although the locations of these complexes do not resemble the symmetric pattern proposed in the simple equations of Johann Heinrich von Thünen and Alfred Weber, they most probably will be geographically dispersed, but to a lesser degree than their markets. Although the level of demand to be covered in these industries is geographically almost equally distributed, there is likely to be some geographical concentration along the lines of the old spatial pattern of GDR industry. This especially applies for the food, drink, and tobacco industry which will be somewhat over-represented in the south of eastern Germany. The localized clusters, however, will barely contribute to the economic regeneration of the eastern German regions: the demand for low qualified labour as well as building materials and agricultural products cannot compensate for the massive loss of industrial skills and production capacities in the eastern German regions produced by the contemporary industrial restructuring.

The second element, represented by investment in the automobile industry, will, in the medium term, also probably result in localized production clusters. However, the tightly integrated supplier networks characteristic of the automobile industry will have nothing in common with the traditional clusters of the construction and the food, drink, and tobacco industries. For investors in the automobile industry, the economic and social *tabula rasa* of eastern Germany provides an almost ideal field for experimenting with the most advanced management practices and production techniques. In the eastern German plants, the European versions of the new management fetish of 'lean production' will be pioneered. Although this concept implies a decentralization of competences and, hence, some local autonomy at the operative level, these plants will be tightly integrated into the pan-European network of production plants controlled by western headquarters. To some extent, this sort of external control will be replicated within the regional supplier network of the eastern German plants: although the demand for flexibility and logistic competence favours local supplier autonomy on the operative level, a considerable share of the eastern German supplier firms will probably remain at the level of second-tier or sub-suppliers belonging to western parent firms which, in turn, are largely dependent on a few large western first-tier suppliers. The emerging regional production complexes will achieve extraordinary levels of productivity, technological flexibility, and quality. However, the dependence of these production complexes on the strategies of a single automobile corporation makes these regions equally extraordinarily vulnerable to external shocks.

The third element, contrasting this future-oriented approach, is composed of the vast majority of western investors, applying simply technologically more advanced versions of familiar Fordist concepts. In combining the Babbage principle with state-of-the-art technology, these investors utilize the eastern German plants as highly specialized mass production enclaves. Since these 'cathedrals in the desert' will not develop regional forward or backward linkages which exceed the modest demand for construction and low level maintenance services, they constitute an impediment for the development of a differentiated regional sectoral structure. The lack of qualified managerial and technical functions within these plants also constitutes a lack of the 'seedbed' which produces entrepreneurs. Finally, their formal rules of work organization do not allow for the emergence of the entrepreneurial skills and 'chaos qualification' which were a feature of the informal networks of the old production system. In fact, the paralysis of these networks as well as the socially integrative functions of the *Kombinate* has been a precondition for the tight integration of the 'cathedrals in the desert' into western corporate networks.

In this sense, the truncated industrialization that has resulted from all three investment strategies amounts to the development of a 'capitalism without (eastern German) capitalists': there seem to be few prospects that investment in the hands of western corporations will trigger self-sustaining regional development. The activities of large western corporations favour a process of geographical integration of eastern Germany, based upon the inclusion of more and more regions into global corporate networks. However, this form of geographical integration should not be interpreted as an indication of an increasingly self-determined participation of eastern German regions in the international economy. With the exception of a few post-Fordist islands in the south of eastern Germany, most will probably remain at the fringe of the international networks of major corporate players. Worse still, with this western truncation of the inherited institutional and social networks, it will take a long time to create the cultural and institutional foundations for a new entrepreneurship which taps into local resources and strengths.

NOTES

The evidence is provided by sixty-eight interviews with managers and union representatives in the chemical; food, drink, and tobacco; metal; and textiles and clothing industries in the eastern German federal state of Brandenburg. These interviews form part of a research project 'Decomposition of *Kombinate* and Regional Development in Eastern Germany' which the author is leading at the Science Centre in Berlin.

1. The proper Granovetter notion is 'undersocialized'.

REFERENCES

Bispinck, R. (1991), 'Collective bargaining in East Germany: Between economic restraints and political regulation', Paper presented at the 13th Annual Conference of the International Working Party on Labour Market Segmentation, Bremen, 11–16 July.

Deppe, R., and Hoss, D. (1989), *Arbeitspolitik im Staatssozialismus. Zwei Varianten: DDR und Ungarn* (Frankfurt on Main: Campus).

Deutsches Institut für Wirtschaftsforschung (1992), *Gesamtwirtschaftliche und unternehmerische Anpassungsprozesse in Ostdeutschland* (Berlin: Fünfter Bericht, DIW-Wochenbericht 12–13/92).

Dicken, P. (1990), 'Transnational corporations and the spatial organization of production: Some theoretical and empirical issues', in A. Shachar and S. Öberg (eds.), *The World Economy and the Spatial Organization of Power* (Aldershot: Avebury), 31–56.

Doleschal, R. (1991), 'Daten und Trends der bundesdeutschen Automobilzuliefer-industrie', in G. H. Mendius and U. Wendeling-Schröder (eds.), *Zulieferer im Netz zwischen Abhängigkeit und Partnerschaft* (Cologne: Bund), 35–63.

Frankfurter Rundschau (1991), 'Aus für Wartburg Produktion', 23 March.

Friedrich-Ebert-Stiftung (1991), 'Rettungsanker Osthandel? Zur Bedeutung der osteuropäischen Exportmärkte für die Unternehmen in den neuen Bundesländern' (Wirtschaftspolitische Diskurse, 25; Bonn: Friedrich-Ebert Foundation).

Grabher, G. (1992), 'Eastern conquista: The truncated industrialisation of east European regions by large west European corporations', in H. Ernste and V. Meier (eds.), *Regional Development and Contemporary Industrial Response: Extending Flexible Specialisation* (London: Belhaven), 219–33.

—— (1993), 'Rediscovering the social in the economics of interfirm relations', in id. (ed.), *The Embedded Firm: On the Socioeconomics of Industrial Networks* (London: Routledge), 1–33.

Granovetter, M. (1985), 'Economic action and social structure: The problem of embeddedness', *American Journal of Sociology*, 91/3: 481–510.

Habermas, J. (1990), *Die nachholende Revolution* (Frankfurt on Main: Suhrkamp).

Handelsblatt (13 December 1990), 'Automobilwerk Eisenach: Schon 1990 werden knapp tausend Vectra gebaut'.

—— (26 September 1991), 'Volkswagen AG: Die ehemaligen Trabi-Werker finden wenig Geschmack an Golf-Montage'.

—— (4 December 1991), 'Philip Morris GmBH: Die ostdeutsche Marke "f6" ist jetzt die drittgrösste deutsche Zigarettenmarke'.

—— (31 December 1991), 'Ernährungsindustrie: Die Strategien der Gross-konzerne zielen über den Euro-Binnenmarkt hinaus'.

—— (2 February 1992), 'Ostprodukte: Zaghafter Vorstoss in westdeutsche Ladenregale'.

Häussermann, H. (1992), 'Regional perspectives of East Germany after unification of the two Germanies', Paper presented at the Conference on 'A New Urban Hierarchy?', University of California, Los Angeles, 23–5 April.

Heidenreich, M. (1992), 'Ostdeutsche Industriebetriebe zwischen Deindustrial-isierung und Modernisierung', in id. (ed.), *Krisen, Kader, Kombinate: Kontinuität und Wandel in ostdeutschen Betrieben* (Berlin: Edition Sigma), 335–65.

Institut der Deutschen Wirtschaft (1991), *IW-Trends* (Cologne: Juni, IW).

Institut für angewandte Wirtschaftsforschung (1990), *Wirtschaftsreport* (Berlin: IAW).

Institut für Wirtschaftsforschung (1991), *IFO-Schnelldienst 16–17* (Munich: IFO).

—— (1992), *IFO Schnelldienst 6* (Munich: IFO).

Lungwitz, R., and Kreissig, V. (1992), 'Sozialer und wirtschaftlicher Wandel in der Automobilindustrie der neuen Bundesländer', in M. Heidenreich (ed.), *Krisen, Kader, Kombinate*, 173–87.

Maretzke, S., and Möller, F. O. (1992), 'Wirtschaftlicher Strukturwandel und regionale Strukturprobleme', *Geographische Rundschau*, 44/3: 154–9.

Marz, L. (1992), 'Dispositionskosten des Transformationsprozesses: Werden mentale Orientierungsnöte zum wirtschaftlichen Problem?', *Aus Politik und Zeitgeschichte*, 24: 3–14.

Massey, D. (1983), 'Industrial restructuring as class restructuring: Productive decentralisation and local uniqueness', *Regional Studies*, 17: 73–89.

Mickler, O., and Walker, B. (1992), 'Die ostdeutsche Automobilindustrie im Prozess der Modernisierung und personellen Anpassung', in M. Heidenreich (ed.), *Krisen, Kader, Kombinate*, 29–45.

Morgan, J. P. (1992), *Investing in Eastern Germany: The Second Year of Unification* (Frankfurt: J. P. Morgan).

OECD (1991), *Wirtschaftsberichte Deutschland* (Paris: OECD).

Schwarz, R. (1991), 'Über Innovationspotentiale und Innovationshemmnisse in der DDR-Wirtschaft' (Discussion Paper FS IV 91–26; Wissenschaftszentrum Berlin für Sozialforschung, Berlin).

Voskamp, V., and Wittke, V. (1990), 'Aus Modernisierungsblockaden werden Abwärtsspiralen—zur Reorganisation von Betrieben und Kombinaten der ehemaligen DDR', *SOFI-Mitteilungen*, 18: 12–30.

Weber, M. (1972), *Wirtschaft und Gesellschaft* (Tübingen: Mohr).

Womack, J. P., Jones, D. T., and Ross, D. (1990), *The Machine that Changed the World* (New York: Harper Perennial).

9

Institutional Change, Cultural Transformation, and Economic Regeneration: Myths and Realities from Europe's Old Industrial Areas

Ray Hudson

INTRODUCTION

The last three decades or so, but especially the last ten to fifteen years, have been a period of continuing decline and deindustrialization in most of those areas that were the original birthplaces of industrial capitalism in western Europe. For many years prior to this, these were locations of not only national but also global significance as centres of capitalist production, associated with a range of industries such as coalmining, chemicals, and metal production and transformation. These old industrial areas were characterized by the employment of waged labour, in big plants owned by big mining and manufacturing companies; associated with this, there was typically a strong gender division of labour between male waged work outside the home and female unwaged domestic labour within it.[1]

The interests of capital were pursued largely unfettered by consideration of those of labour in the initial phases of establishing capitalist social relations in these areas. With the passage of time, the dominant capitalist firms developed institutional forms and linkages through which their interests could be organized and relationships between them regulated. In this way, as well as through political and social relationships developed within the sphere of civil society, the emergent class of industrial capitalists formed itself in these areas (see Grabher 1990; Hudson 1989). Over the subsequent years, too, slowly, haltingly, and often as a result of bitter capital–labour conflict, an embryonic working class began to establish itself in its emergent places and institutional arrangements were constructed to represent the interests of workers and their families counterposed to those of capital (see e.g. Thompson 1968). In part, this involved the creation of trade unions to represent the interests of labour as wage-labour *vis-à-vis* capital in the workplace, characteristically associated with ongoing struggles over the conditions, terms, and wages associated with work. In part, and generally linked to this, it involved the formation of political parties to represent more general working-class interests, both as producers and also as consumers, often articulated through claims for state provision of housing, educational, health, and social services. Indeed, typically over the years there was a growing national state involvement in regulating the conditions of reproduction of labour power in such areas and industries and, on occasion, those of production itself. In this way, national states established regulatory frameworks that mediated relationships between local areas and an international economy and set the limits within which local institutional forms have been created (see e.g. Lash and Urry 1987). A history of dependency upon wage-labour and service provision by major private-sector industrial enterprises often resulted in an easy slippage to dependence upon the state as a source of jobs and services. Indeed, the balance of private and public-sector involvement in the provision of jobs, housing, and services has varied precisely because such a switch was often an explicit objective of working-class political parties and trade unions in these old industrial areas.

The net result of this history was the creation and reproduction of a culture of dependency in these old industrial areas, in three senses. First, upon waged labour provided through major capitalist enterprises and/or the state; secondly, upon the provision of housing and services through these same channels; thirdly, upon ways of understanding the world, and possibilities for changing it, that were profoundly shaped by these particular forms of social relations of production and consumption. This culture of dependency was thus sustained through the particular and thick institutional tissue of such areas and through widely accepted beliefs as to what was possible by way of change and via practices as to what was regarded as appropriate behaviour within such areas.[2] The sense of social order conferred by trade unions within the workplace diffused more widely through the institutions of local civil society in such areas, but within the context of a hegemonic culture of waged labour. As a result, the often profoundly paternalistic social relationships within the workplace became extended into mechanisms for social ordering and control in the community beyond the workplace, transmitted and communicated through the institutions of local civil society and/or the state. In this way, a social fabric of community evolved, with its constitutive institutional tissue, through which people came to know and understand their area, the world around them, and their place within it as wage workers or dependants of wage workers.

As long as effective demand for commodities produced in and through such areas remained high—or at least exhibited only cyclical fluctuations rather than secular decline—the economic, social, and political basis of such areas could be reproduced, albeit within the defining parameters of capitalist relations of production. From the late 1950s, however, it became increasingly apparent that secular decline rather than cyclical fluctuation was becoming the order of the day. An emergent new international division of labour and global shifts in geographies of production was accompanied by deindustrialization, capacity cuts, job losses, and decline in old industrial areas that had once been at the forefront of the process of capitalist industrialization (see e.g. Dicken 1990; Hudson and Sadler 1989; Sadler 1992). From the mid-1970s the pace of secular decline accelerated and came to characterize a widening range of industries (see e.g. Mandel 1978). This raised important theoretical questions in terms of how to understand such changes and also practical ones in terms of what might be done to reconstruct an economically viable future for such areas. Prima facie, for example, it seems unlikely that such old industrial areas, with their particular heritage, would be feasible locations for growth strategies based around expanding activities such as professionally based knowledge—intensive business services. The particular *types* of thick institutional structure that had been evolved within them were simply not appropriate, repelling rather than attracting such activities. What sorts of activities could therefore feasibly provide a new economic basis for such areas? What sort of cultural and institutional changes might be necessary within such areas to enable them to reposition themselves more advantageously within a global political economy? To what extent are such necessary changes possible? To transpose a point made by Offe (1975) into a slightly different context, to what extent are necessary cultural and institutional changes impossible, impossible cultural and institutional changes necessary if a viable alternative economic basis is to be constructed for these old industrial areas?

There are, however, two further and crucial points. First, even if new institutional forms are necessary for economic regeneration, it by no means follows that they will be sufficient for it (see also Martin 1988). Secondly, if the old forms of localized institutional structures were to be deconstructed and new ones successfully created in their place in pursuit of this goal of economic regeneration, which class interests and social groups' interests would they favour? What, in this sense, would be the social context and content of a re-created locally specific institutional tissue? The remainder of this chapter considers a number of alternative solutions proposed for such areas, a range of options that must be understood in relation to the limits set by the national regulatory frameworks as they themselves have been radically altered over the last fifteen years or so in most of the advanced capitalist states of Europe. In addition, it also examines the extent to which these presuppose particular localized forms of cultural change and involve innovations in institutional arrangements. Some comments will also be made about the feasibility of these varying options, identifying the implications of this for the future of these areas.

PRODUCTIONIST SOLUTIONS

Small and Medium-sized Manufacturing Firms and the Enterprise Culture

The social-class legacy of a dependence upon wage-labour in large plants of major companies has often been seen as a significant cultural barrier to the economic transformation of the old industrial areas via the fostering of an enterprise culture, centred around new small firms and self-employment (see e.g. Sengenberger and Loveman 1987).

From the late 1970s, however, as many national governments retreated from direct confrontation with problems of spatial inequality, one strand of policy saw the emphasis in urban and regional regeneration shift dramatically to approaches that centred on endogenous growth of new small firms in problem areas.[3] For such small-firm policies to be successful presupposed, *inter alia*, a decisive transformation within the dominant working-class cultures of old industrial areas to embrace the values of entrepreneurship, self-employment, and the enterprise culture. A key intervening variable in producing this cultural change in attitudes and practices was seen to be the creation of new mechanisms and institutions to nurture the formation and growth of new small companies; *a fortiori*, to persuade ex-wage workers to become self-employed or, better still, to become themselves employers of wage-labour. To this end, local governments had to acquire new skills in place promotion and in formulating local economic development measures while a plethora of new institutions, often bridging the private- and public-sector divide (such as local enterprise agencies) were established in old industrial areas to help create the conditions in which new small firms could be born and then grow.

An emblematic example of these processes of change is provided by the town of Consett and the surrounding Derwentside District in north-east England (see Hudson and Sadler 1991). For almost a century and a half, the District's economy and society had revolved around the iron- and steelworks and coalmines of the Consett Iron Company. The company built the settlements of the District to house its workers and its deeply paternalistic influence penetrated local life far beyond the workplace. Over time, with growing state involvement in housing and service provision, and the nationalization of the coal and steel industries, this became transformed into a deep dependence on a statist form of Labour party politics that was later rudely challenged by the closure policies of the nationalized coal and steel industries. Thus the way was opened for an attempt radically to transform the area's economy and the dominant cultural legacy of its industrial history. The much-publicized closure in 1980 of the British Steel Corporation's Consett works symbolized the deindustrialization of the

District, although in fact this had been preceded by over three decades of industrial decline there. Following the steelworks closure, Consett became the focus of a vigorous campaign to create a newly diversified local economy, to replace the jobs lost in large plants with new ones born of a flourishing enterprise culture in new, local small firms. This reindustrialization policy required radically different local institutional arrangements, involving cooperation between local (Labour-controlled) government, central (Conservative-controlled) government, new agencies such as British Steel (Industry) and British Coal (Enterprises), subsidiaries of the nationalized industries, and the private sector. These institutional innovations were represented most clearly in the creation of the Derwentside Industrial Development Agency (DIDA), a local enterprise agency launched in a blaze of publicity and intended to facilitate the emergence of this new economy. There is no doubt that in the 1980s Derwentside witnessed the vigorously publicized promotion of a 'reindustrialization strategy', centred around the existence of 'probably the best project package in the UK' (DIDA and DDC, n.d.). Thus, whatever the new institutional arrangements in Derwentside, the key point was that they sought to promote an enterprise culture via creating a (temporarily) supportive and highly state-subsidized environment for potential entrepreneurs. DIDA took a high profile role in publicizing this fact and in stressing its involvement in the preparation of Business Plans for potential new businesses.

At one level, it would seem that institutional innovation proved highly successful in bringing about the desired local economic transformation. According to DIDA, £50 m of public expenditure was undertaken in Derwentside District between 1980 and 1988. This in turn triggered over £70 m of private-sector investment, involving more than 200 businesses. Moreover, by 1989 it was claimed that this had resulted in the creation of 4,500 new jobs, mostly in a diverse range of indigenous, locally owned and controlled, competitive, and profitable new companies. Such claims, however, have to be understood not so much as a description of what has happened but as part of a process of place promotion and, in some instances, institutional self-justification. For those involved in the new institutions have a vested interest in promoting their efficacy and the success of the policies that they promote and implement.

In contrast to DIDA's claims, other data, from central government's own *Censuses of Employment* suggest a much more modest net growth of 1,270 manufacturing jobs in Derwentside between 1981 and 1989, compared to a decline of over 8,000 between 1978 and 1981. Other data from the District Council's own 1988 *Progress Review* suggest considerable volatility of employment, with a marked turnover of births and deaths of new firms (see Hudson and Sadler 1991). Moreover, it is highly debatable as to whether many of the new firms, especially the relatively larger and/or more successful and well-publicized ones, can be appropriately described as 'local'. Most of these were established not by 'local heroes' but by 'entrepreneurial immigrants' (Caulkins 1992), attracted by grants, loans, and the mass of unemployed people desperate for waged work. Indeed, it is richly ironic that by far the most successful 'local hero', Bob Young, is an ex-coalminer whose company, R and A Young, has its core activities in opencast coalmining. The old coal economy apparently remains the most fertile breeding ground for such a new enterprise culture as may be emerging in Derwentside!

There is, then, little evidence that the new institutional arrangements put in place in Derwentside in the 1980s have had anything more than a cosmetic and superficial effect in bringing about a cultural transformation there that embraces the values of entrepreneurship and enterprise. There is little evidence of local people setting up their own companies; those few that have done so have typically been motivated by a defensive fear of unemployment rather than an offensive embracing of the enterprise culture. Far from being unique in this, Derwentside is simply one of many examples of European old industrial areas in which the legacy of a culture of waged labour has proved highly resistant to change, though sometimes because unemployed former workers recognize only too well that the economic climate in the areas in which they live is a far-from-promising one for the would-be entrepreneur (see e.g. Rees and Thomas 1989). Moreover, most of such new small firms as are established are in low tech services rather than high tech manufacturing (see Storey 1990; Hudson *et al.* 1992). And for those who do set up their own small companies, 'enterprise means . . . a twilight world of hard work, low pay, casual labour and insecurity as . . . people plod along trying to secure a decent living through enterprise' (MacDonald 1991: 3). Once again, this suggests that the prospects for an emergent enterprise culture are less a matter of individual psychology or institutional innovation than a consequence of economic depression, and precarious labour-market conditions in old industrial areas.

It is certainly true that in some old industrial areas there is evidence of burgeoning small-firm formation, though there is much less evidence of a re-creation of the regional economy through Marshallian industrial districts of linked small firms. In Germany, for example, almost 12,000 new companies were registered in North-Rhine Westphalia (NRW), which includes the Ruhr Area, in the first six months of 1991 alone, some 25% of the national total (Parkes 1991). But such success stories only emphasize the importance of the national and local economic climate and the national regulatory framework and public expenditure policies on education and training. Before 1965 there was not a single university in the Ruhr Area. There are now six, so that the region currently has one of the densest university landscapes in Europe. There are also six polytechnics, eleven

technology centres, four centres of the Max Planck Institute, and two
Fraunhofer centres (Hassink 1992: 55–6). Moreover, major private-sector
companies, coming together as the Initiativskreis Ruhrgebiet in 1989, have
so far invested 5.0 bn DM, committed another 4.5 bn, and earmarked a
further 12 bn DM expenditure on industry and infrastructure in the Ruhr
by 1996. In this way, established major private-sector interests in the area
are stepping in to fill a gap left by the switching of federal government
expenditures to the territory of the former GDR, thereby sustaining an
economic climate in the old industrial area of the Ruhr that encourages the
formation of new small firms.

It is also true that new institutions intended to foster the emergence of an
enterprise culture in old industrial areas have had some effects as intended,
albeit effects that are limited by the combined impacts of their cultural
history and the contemporary politico-economic environment in which
they exist. Such institutional innovation is, however, of much less
significance than the investment and spending policies of big private-sector
companies and central government in such areas in creating and sustaining
an economic environment supportive of small-firm formation and growth.[4]
Whilst the economic impacts of institutional innovation may be limited, the
greater significance of such changes may well be—and be intended to be—
ideological and political, seeking to create the illusion that an enterprise
culture can be conjured up in even the most unpropitious of circumstances.
For if it can be seen to be created in old industrial areas, then surely it can
be created anywhere.

Big Firms and the Branch-Plant Economy

The enterprise culture project essentially aimed to replace large firms
with small ones as a source of wage labour and/or waged labour with self-
employment, or even emergent capitalist status, for some people as part of
a radical class-restructuring of old industrial areas. This implies a
simultaneous reworking of the boundaries and composition of the working
class and the creation of a new fraction of the old middle class of petty
capitalists. An alternative solution, with different implications in terms of
cultural and institutional change, is to seek to create a new economy
around the branch plants of multinational companies but in different
industries from the old ones and employing wage labour within a reformed
working class on very different terms and conditions to those 'traditionally'
found in the old industries. Thus industries that are 'new' to these 'old'
areas, which have become characterized by high unemployment, are
attracted by their supplies of 'green' labour from which they can carefully
select their workforces as the route to trouble-free production, and high
levels of labour productivity and profits. This involves not so much an
abolition of the working class but rather a radical reformation of it in terms

of its age and gender composition and attitudes towards work and trade-unionism.

However, to attract such multinational branch-plant investment in a global place market requires new regional and local institutions to promote the virtues of areas as spaces for profitable production. In part, this has been linked to a territorial decentralization of powers within some national states so that the creation of new promotional institutions is itself part of a restructuring of the national state; in other cases, such institutions have been grafted on to existing state structures, often in an *ad hoc* way. But in both instances the intention is to create new institutions that can act as 'one stop shops', selling the attractions of areas to multinational companies in terms of the availability of grants and loans; the quality of the built, social, and natural environments; and, often crucially, the adaptability, flexibility, and passivity of the labour force. These new promotional institutions typically take a pseudo-corporatist form, with carefully selected trade union representatives who promote the virtues of the area's workers, usually in the hope of securing the right of representing them in new single-union deals.

A typical example of these sorts of institutional changes can be found in north-east England, where the prime role of the Northern Development Company, established in the 1980s as the latest in a line of such agencies, has been to promote the attractions of the 'Great North' to multinational companies, especially those from Japan (see Hudson 1991). The NDC and its predecessors have played an important role in persuading Japanese companies such as Fujitsu and Nissan to invest in branch plants in the north-east in fierce competition with other areas within the United Kingdom (Dicken 1990; Spellman 1991), elsewhere in Europe and, indeed, in the world. The process of inter-area competition is a pervasive one and this in itself creates pressures for innovations in the institutional structures through which old industrial areas are sold—or sell themselves—to multinational capital.

Now, in many respects, this contemporary pursuit of multinational investment is similar to policy responses to the onset of industrial decline in old industrial areas from the late 1950s. Then, too, the attraction of branch plants of multinationals, above all those based in the United States, was seen as central in creating an alternative economic basis for these areas. In addition, attracting such new inward investment was and is seen as something that required institutional changes to secure the conditions that would entice it to old industrial areas. In the United Kingdom in the 1960s, for example, this involved the creation of new corporatist Regional Economic Planning Councils and a restructuring of local authorities and a redefinition of their role *vis-à-vis* local economic development. Equally, it was seen as necessitating profound changes in working-class culture, both in terms of trade-union attitudes towards industrial relations and in terms of lifestyles. But the pursuit of such policies in the 1960s and 1970s created, for some, fears of external control, vulnerability, and dependency in a branch-plant economy (Firn 1975) as formerly coherent regional economies became 'global outposts' within corporate production chains (Austrin and Beynon 1979). Moreover, it was precisely the failure of such policies to deliver sufficient new jobs to old industrial areas that helped open up the space in which very different policies of self-reliance, endogenous growth, and the encouragement of small firms could evolve. These policies in turn failed to create even a fraction of the necessary new jobs, hence reopening space for the renewed emphasis on attracting multinational investment and on constructing new regional and local institutions to enable old industrial areas to compete more effectively for it. Hence, such areas embarked on further circuits around the mulberry bush.

In other respects, however, the pursuit of multinational investment in the 1980s takes place in a very different context from that of the 1960s. For, whilst it is unlikely that the regional economy will be reconstructed around spatially agglomerated complexes of companies linked into just-in-time production systems (see Mair 1991), there is evidence that multinational companies are now pursuing more 'regionalized' versions of their global strategies, with an organization of R&D, production, distribution, and sales on a, say, European rather than worldwide basis. Clearly, if branch plants cease to be the 'global outposts' of transnational production chains but become more embedded (see Chapter 2, above), in a variety of ways, into local and regional economies, the potential for stimulating local economic development may be greatly enhanced. This raises questions as to what sort of state policies may be necessary to attract such investments, with a much greater concern for quality in terms of the environment, labour-force skills, and so on. However, whilst such embedded branch-plant investment may provide a greater economic coherence at one spatial scale, it does not necessarily remove the uncertainty surrounding the future of particular old industrial areas. This uncertainty in turn increases the pressure for institutional innovation and to invent fresh competitive advantages in marketing one area against others. Moreover, against a background of high unemployment, multinational companies can exercise great selectivity in whom they choose to employ, creating new divisions within a redefined working class. Such companies effectively exclude the vast majority of people in such areas as part of their potential workforces. Nevertheless, it may well be the case that attracting major branch plants, especially those which concentrate on the more technically sophisticated, high value-added segments of the production chain, offers the best opportunities for some employment creation in many old industrial areas. If this seems a pessimistic conclusion, it may also be a realistic one.

CONSUMPTIONIST SOLUTIONS

From Working-Class Production Spaces to Tourism based on the Heritage of Working-Class Production

Given the generalized switch towards a tertiarized economy, it is no surprise that in some deindustrialized old industrial areas, attempts have been made to identify potential service sector activities that might be developed there. Nor, given the environmental characteristics of such areas, often blighted by the scars of industrial production followed by decay, is it any great surprise that in some of these areas attempts have been made to transform the legacy of earlier industrial production into the basis for contemporary employment and economic activity (Hudson and Townsend 1992). Thus the heritage of deindustrialization is not to be completely cleared away as part of the process of seeking to attract fresh industrial investment but is to be selectively preserved and transformed. The past is to be restored in a partial and sanitized form as a tourist attraction for people who wish nostalgically to engage with a typically romanticized version of an earlier industrial era. Indeed, whilst in some cases it is the preservation or reconstruction of actual industrial landscapes, of old ways of work and living, that is the focal point of attraction (as in Bradford or in museums such as Beamish in north-east England or Big Pit in south Wales), in others the selling point is a fictional world that itself portrayed life in the industrial past through rose-tinted lenses (as, for example, south Tyneside becoming Catherine Cookson country). In short, this is the old industrial areas' particular version of the heritage industry (Hewison 1987).

To represent old industrial areas in this new guise again has necessitated important changes in local and regional institutional arrangements. New regional Tourist Boards have been created and local-government economic-development departments have had to develop new skills and expertise so as to invent and then market the attractions of their areas. Such heritage sites have often been linked to the formation and growth of local small-scale service activities, such as pubs, clubs, and hotels. In these ways the tourist industry is again associated with important gender and age changes in the composition of the local working class, with a switch from full-time to part-time and casualized precarious work, much of which is non-unionized (Hudson *et al.* 1992). Such radical changes both presuppose and help bring about significant changes in local culture, in the understanding of what waged work means. For working for a wage via serving other people carries with it very different connotations from working for a wage via cutting coal or producing steel or ships. Areas that were formerly centres of industrial production become economically dependent upon incomes earned elsewhere, both locally and more widely. However profound the changes in local culture, and in attitudes to work, and however innovative the marketing strategies of promotional institutions, tourism remains a profoundly risky basis on which to rebuild a local economy, subject to the whims of tourists as much as to the effects of economic crises in other areas.

There is no doubt that the turn to tourism in many old industrial areas represents a sort of politics of despair, of a lack of alternatives in terms of manufacturing industry or other types of service-sector activity. It is a recognition of economic marginalization in areas that were once focal points in an accumulation process that now increasingly by-passes them, with deep implications for class structures and social relationships within them. There may be circumstances in which promotion of such activities might appropriately form one element in local economic-development strategies but to rely upon them as their main, or sole, basis is a very risky course of action. For it places such areas in competition with the vast array of locations, scattered around the globe, which seek to sell themselves to tourists in various ways. There is no guarantee that a tourism based on nostalgia for a departed industrial past will prove more attractive to consumers than one based on sun, sea, and sand in southern Europe or the mystic delights of distant continents as the tourist industry becomes increasingly globalized.

From Working-Class Production Spaces to Middle-Class Residential and Consumption Spaces

The enterprise culture, the branch-plant economy, and the turn to tourism, whilst in many ways drastically differing, do share one common characteristic; they are all strategies that seek to employ people in an area as waged labour, as workers involved in the production of goods or provision of services as commodities. Certainly these various options imply drastic local labour-market restructuring, via age, gender, and so on, with deep cultural change in norms as to the meaning of work, working conditions and practices, and lifestyle and living conditions. But at least they were predicated upon capital's interests in these old industrial areas as locations for the production of profits, in at least some of their populations as wage workers. In this sense, they all presuppose the reproduction of a working-class culture of wage-labour, albeit of different kinds.

In other old industrial areas, there have been radically different responses to deindustrialization, decline, and decay. Above all, these hinge around a profound and selective demographic and social recomposition as, via a mixture of gentrification and comprehensive physical redevelopment, these areas are reworked as sites of middle-class residence and consumption. This process is most sharply symbolized by waterside locations within the old conurbations, but it is by no means confined to them. Nevertheless, the visual images of former areas of derelict docklands, warehouses, shipyards, and associated working-class housing areas converted by a combination of refurbished former buildings or brand-new residential ones on sites cleared of their former occupants, occupied by the new middle classes and surrounded by equally new retailing and leisure complexes targeted at meeting their demands and those of the more affluent fractions of the working class, are very powerful ones. There are sometimes passing gestures to 'high tech' production, but these are essentially cosmetic, for the rationale of these new complexes, shimmering with reflective glass and adorned with the trappings of postmodern architecture, is consumption. The dominant themes are those of visibly conspicuous consumption, social recomposition, and political change, physically symbolized in the contrast between the old and new built forms.

These are very dramatic transformations and have typically required very drastic institutional changes to facilitate them. Above all in the United Kingdom, these have been identified with the establishment in the 1980s of the Urban Development Corporations, non-elected bodies appointed by and generously funded by central government. The UDCs are a type of authoritarian corporatist institution, driven by the interests of particular sections of the private sector with token representation of the interests of organized labour and local residents. Local Authorities, typically Labour-controlled, in areas where UDCs operate have tended to accept the situation, albeit unwillingly, and work with them, not least because they represent a substantial source of extra central government expenditure in an era when this has generally been shrinking. They have a brief to transform selected old industrial areas within the conurbations via a speculative, private-property-led redevelopment strategy. The task for the UDCs is to cut through the regulatory controls formerly exercised by elected local authorities and to provide an environment in which private capital will invest in spectacular redevelopment projects of the desired type, deflecting local criticism by claims that such projects will result in the creation of thousands of jobs in the midst of areas blighted by high unemployment and widespread poverty.

The activities of the Teesside Development Corporation are by no means untypical of the UDCs outside of London.[5] The TDC sees its mission, in the words of its Chief Executive, as 'market led and about creating jobs and confidence' and its portfolio of projects reveals starkly what this means in practice. They include a £10 m offshore technology park on the site of a former shipyard; a £40 m marina, residential, and leisure complex in the old Hartlepool coal docks; and an £80 m project to convert the derelict Stockton racecourse into a leisure and retail centre. The precise extent to which, and the forms in which, these proposals will come to fruition will depend upon private-sector investment decisions.

Equally, whilst there is considerable uncertainty about the precise employment figures and types of jobs, suggestions that all this will create 15,000–20,000 jobs help defuse local opposition and persuade local authorities in the area to try to work with, rather than against, the TDC. But there is no doubt that it is the TDC and not local authorities that sets the agenda for development, for what is to be defined as development, on Teesside in the 1990s: private-sector-led speculative property redevelopment resulting in increasingly sharp social and spatial differentiation within a post-industrial, post-production Teesside, juxtaposing middle-class affluence with working-class poverty and unemployment. In so far as such investment generates new jobs, they are typically part-time or casualized, unskilled, poorly paid, and non-unionized.

The sort of future being created on Teesside is far from unique, as the other UDCs pursue similar sorts of policies. In other areas, different institutional mechanisms have been devised to pursue the same objectives, but all are essentially derivative of a US-based urban redevelopment model imported into Europe (see e.g. Judd and Parkinson 1990). The recent (1991) explosion of riots in some English inner cities suggests that the effects of such visibly divisive policies may well not be consistent with the establishment of a stable localized mode of regulation in such areas.

THE WELFARE STATE SOLUTION: FROM INDUSTRIAL WORKERS TO CLIENTS OF THE WELFARE STATE

For some deindustrialized old industrial areas, there is no foreseeable future role as locations of production, no feasible possibility of reconversion into places of middle-class residence and consumption. For them, the only foreseeable future is one of dependence upon state transfer payments as the vast majority of their populations become clients of the welfare state that typically has been drastically cut back in scope.

In one sense, this is not new. There have long been people dependent upon such transfer payments. What is new, and especially stark in former one-industry settlements, is the generalized and simultaneous occurrence of mass unemployment in many such places. This in turn has required new forms of institutional arrangements, new forms of state organization specifically focused upon social containment and control via varied training, retraining, and temporary employment schemes. In the United Kingdom, this has been most clearly manifest in the activities of the Department of Employment/Manpower Services Commission and schemes such as the Youth Training Scheme, the Community Programme, and Employment Training. These new national schemes and state organizations have come to play a pivotal role in social regulation in many old industrial areas, centred on the allocation of places on training and temporary employment schemes and the management of a substantial surplus population. In Derwentside District in the late 1980s, for example, over 2,800 were engaged in such schemes. To put such figures into context there were around 14,000 people formally employed in the District at this time and around 3,500 registered as unemployed. Of the latter, 40% had been continuously unemployed for twelve or more months, 25% for twenty-four or more months, with the recorded incidence of such long-term unemployment depressed by the impact of temporary job schemes. As a consequence of the low levels of welfare payments, there has often been a parallel growth in the black economy (in Cleveland County perhaps one in ten of those registered as unemployed are involved in the black economy, for example) or even in criminal activities in these areas (as became clear in the Meadow Well riots in 1991 on Tyneside) as people seek to supplement inadequate incomes from transfer payments in a situation in which they are effectively excluded from the formal labour-market. In old industrial areas in other parts of western Europe, the limited citizenship rights of temporary—or, still more so, illegal—migrant workers both limit their welfare rights and exacerbate the pressures pushing them into black work or criminal activity.

As Pugliese (1985) some years ago wrote of much of the population of the Mezzogiorno becoming clients of the welfare state, so too can one now write of a very substantial fraction of the populations of many old industrial areas of the United Kingdom in these terms. Such people are often consigned to the worst of the remaining stock of public-sector social housing, and are subjected to high levels of surveillance by police and social security officials, especially so in those locations in which they are spatially adjacent to the transformed consumption spaces of the new middle classes. Once again, such a pattern of sharp socio-spatial segregation is one that is increasingly common within western Europe's old industrial areas.

CONCLUSIONS

The five scenarios or alternative futures for old industrial areas are put forward here as analytically distinct options. In reality, elements of more than one of these will be found in particular times and places, interlinked in practice as the uneven development of capital is reproduced in new forms. Moreover, whilst they have been presented here as analytically distinctive options, they must be understood as linked together as elements of, and experiments within, a wider political vision of neoliberal policies, predicated upon selectively cutting back the state and promoting competition and the market as the main mechanism of resource allocation within society.[6] As central governments aggravated, if not provoked, local economic crises in old industrial areas via their macro-economic, monetary, and fiscal policies, they simultaneously withdrew from serious engagement with their resulting localized and regionalized consequences. Rather they used these as a further opportunity to promote their own political projects via new policies which sought to encourage competitive localized solutions to such problems; and, moreover, they sought to secure generalized acceptance of a view that there was no alternative way of tackling such problems. The result was a series of experiments at local level, varied in form and content, but unified by their origins in broader national, and at times supranational, political projects as Thatcherism and Reaganomics, for a time, sought a global hegemony.

Nevertheless the fact that there is a range of possibilities within a particular matrix of opportunities, albeit one that is far from optimal for old industrial areas, raises important practical questions concerning their future. These areas had their origins as workshops of the world, as gateways to the world. They were significant centres of capitalist production and trade, with characteristic social class structures, institutional and cultural traditions, and practices associated with particular mixes of industries. And now? The cohesion that these areas formerly had is collapsing or has collapsed. They are increasingly characterized by internal fragmentation and dislocation as what had previously been relatively coherent—within the limits set by capitalist social relations—local and regional economies shatter and local and regional societies are subject to intense pressures under which they threaten to rupture, perhaps irrevocably.

How then can such areas position themselves, or be positioned, more favourably in the context of: a rapidly shifting global economy; further deepening of integration within and widening of the European Community; the failing capacities of national states to regulate social and economic life and maintain socio-spatial inequalities within 'acceptable' limits? What are the most appropriate and realistic objectives for such areas—and what are the most appropriate institutional arrangements at global, European Community, national, and subnational levels to try to ensure that these are attained in an uncertain and fluctuating global political economy? As the preceding pages indicate, despite often intense local efforts and innovativeness, aimed at creating fresh new localized and thick institutional tissues, many of the 'solutions' attempted so far have been at best partially successful. Is it, then, the case that the institutionally necessary really is impossible and the institutionally impossible necessary? It would, on the evidence presented so far, be easy to conclude, echoing voices heard in the Depression of the 1920s and 1930s, that in fact there is no solution for the problems of such areas, bar complete abandonment. It should certainly be emphasized that there are no 'local' solutions to 'local' problems and no amount of localized institutional and cultural change *on its own* will guarantee local economic prosperity. Local institutional innovation and local proactivity can never in this sense be *sufficient* to guarantee economic regeneration and social progress. The real questions are much more to do with the relationships between local changes, corporate strategies, and national and supranational state strategies and with *whose* solutions are to be implemented in such old industrial areas.

In this context, it is important to remember that the localized thick institutional structures that evolved in the past have often become a mechanism to stifle dissent or hinder opposition to what was regarded as the conventional wisdom of orthodox solutions. During the 1960s in north-east England, for example, militant miners who wanted to take industrial action against colliery closures were threatened with disciplinary action by their own trade union because it would embarrass a Labour government that was closely identified with the organized labour movement (Douglass and Krieger 1983). More generally, attempts to contest plant closures and job losses have often been hampered by inter-union, or even intra-union, disputes (see e.g. Hudson and Sadler 1983; 1986). Moreover, as the 1984–5 miners' strike in the United Kingdom dramatically demonstrated, regional chauvinism and territorial division within trade unions can decisively influence the trajectory of economic and social change within different areas built up around the same industry (Beynon 1985). In these circumstances, the organizational structure of labour itself becomes a barrier to effectively contesting, and posing a credible alternative to, the futures proposed for particular local areas by capital and/or national states. During the 1980s attempts to establish grass-roots local community-based solutions to the economic and social problems facing Derwentside in the wake of the closure of Consett steelworks were effectively marginalized as the local Labour-dominated District and County Councils laid heavy emphasis upon a conventional reindustrialization programme. Resources were channelled into this through the state whilst they were simultaneously denied to local grass-roots initiatives. This was by no means a process unique to Derwentside. In all these cases, albeit in different ways, the legacy and residue of a former thick localized institutional tissue suppressed the exploration of alternatives to, and resistance to, the conventional solutions. Furthermore, it is not just the legacy of the past that stifles the exploration of alternative futures for such areas but also the way in which attempts are made to create new institutional structures which will provide part of a regulatory framework for new trajectories of local economic development. For example, the incorporation of key individuals in the organized labour movement into territorially based coalitions that seek either to seduce the branch plants of transnational capital and/or promote the growth of small firms and a localized enterprise culture serves to define an agenda which excludes more radical, alternative policies in and for such areas. Under these circumstances, it would seem that localized institutional thinness may have held greater emancipatory and radical transformatory potential.

There is a further point. It would seem that particular combinations of local economic growth models and localized institutional structures, which play an important regulatory role, evolve in such a way as to provide a stable localized basis for production and social reproduction. Whilst this stability may be sustained for longer or shorter periods of time, in the end serious disjunctions emerge between the growth model and the regulatory framework and a localized crisis erupts. This is resolved—or more accurately, attempts are made to resolve it—via searching for a new stable combination. What remains an open question is whether localized institutional structures can be developed which would allow more of a smooth and incremental adaptation of local economic and social life to the broader exigencies of national state policies and global political–economic change. The evidence to date suggests that this is unlikely, at least in the vast majority of old industrial or deindustrialized areas.

What then can we conclude? Perhaps the real task is to accept that as long as there is capitalism there will be uneven development, and so to seek to find the most appropriate strategies within these limits to allow people to live, learn, and maybe even work in the old industrial areas in which they feel most comfortable. But, in seeking to define such strategies, there is a deep ambivalence surrounding the role of local institutions. Certainly, there are no simple relationships between the existence of local institutional thickness, guarantees of local economic regeneration and growth, and the socially progressive and transformatory content of the politics pursued through it. Rather, the relationships between localized institutional structures and localized economic and social change are ones that both are reciprocal and vary between places and over time. Local institutions may be important in some circumstances in fostering change, in others in resisting it. Whether encouraging or opposing change is politically progressive or regressive is, however, another matter. There are those who place great faith in a constitutional reform and decentralization of power to regions as the way of furthering working-class interests in old industrial areas (see Byrne 1990). This is to adopt a very partial view of the issues. In other circumstances, local institutions may be, at best, of marginal significance. The key theoretical questions are, therefore, to do with understanding this complex interplay between those circumstances in which local institutional structures do not or do matter, the ways in which they matter, and the political implications of their influence. The key practical questions are then to do with specifying the most beneficial—or least damaging—local strategies in the light of this knowledge.

But what it does seem reasonable to say is that there can be no localized solution that impacts equally on all social classes and groups in an area. In this sense, the myths of socially undifferentiated proactive localities seeking to pursue their own interests are dangerously misleading ones. What it is necessary to do is specify *whose* interests are to be prioritized in seeking to formulate and implement local solutions and create appropriate localized institutional structures to facilitate this process. If the intention is to promote a selective social and political recomposition via transforming old industrial areas into new middle-class consumption spaces, or into islands of an enterprise culture, then it must be recognized that such solutions are unavoidably locally divisive as many more people will be excluded from than will be included in these particular projects.

Clearly, the welfare state solution is more inclusive but it is premised on accepting marginalization, on accepting generalized expulsion from the working class to the ranks of the surplus population. Whilst there are clear dangers in fetishing wage-labour, there are at least equally serious ones in passively accepting the poverty, poor living conditions, and ill-health that follow from prolonged unemployment. In this sense it ought to be challenged, consistently and strongly. There are those who suggest that organizing marginalized social groups so as to make increasing demands for improved living conditions via enhanced state transfer payments offers a route to a more generalized progressive transformatory politics (Byrne 1990). Whilst there is a very strong case for increasing the quality of life of the members of such groups, it is unlikely that this would be translated into a wider radical reformism.

Solutions which are predicated upon a recomposition of the working class are certainly more inclusive, which is not to deny they are also divisive. Both in terms of economics and politics, a future based on commodification of the remains of an industrial past for present-day tourists is less attractive than one based around new forms of production in the branch plants of new industries. Nevertheless, at least in the old industrial areas, in an era of increasing globalization this sort of local economic reconstruction via inward investment would require deep and radical policy changes at national and supranational level as well as local institutional innovation. It would, however, have to be acknowledged that this sort of reindustrialization policy would offer little chance of a return to 'full employment' in the local economy. The ability carefully to select workers against a background of high unemployment is typically a decisive element in such investment decisions. However, policies to improve the quality of the local productive environment, not least in terms of a more skilled and qualified workforce, would increase the chances of attracting factories that would require a wider range of occupational skills and, maybe, greater local multiplier effects. Given that such reindustrialization policies will not lead to 'full employment', however, there will be powerful need for strong welfare policies to provide for those who will remain unemployed, and this will require decisive changes in national and perhaps supranational policies. If this seems like a pessimistic conclusion, postulating a scenario that, even if it came about, would offer only a limited, and far from ideal, solution, then this is no more than a reflection of the limited choices open to most people in the old industrial areas faced with the realities of globalization and the structural constraints of the contemporary political economy of capitalism.

NOTES

This chapter draws upon a number of years' research into the causes and consequences of industrial restructuring in old industrial areas within western Europe. In particular, it draws upon research in northern Britain but reference to other areas is made as appropriate.

1. In some areas and industries, such as the textile districts of northern England, women were of greater significance as industrial waged workers and this had important ramifications for gender divisions of labour, institutional forms of representation of working-class interests, and social relationships within civil society (see e.g. Massey and McDowell 1984). In general, however, the stereotypical pattern of waged male labour outside the home and unwaged female labour within it was dominant in these old industrial areas.

2. The concept of an 'institutional tissue' refers to the existence of institutions both in the spheres of the state (for example, local government) and civil society (for example, trade unions and political parties). Indeed, one of the characteristic features of old industrial areas, typically under the sway of social democratic politics, is that the boundaries between state and civil society were both blurred and permeable. Not least, key individuals typically were active in institutions on both sides of the state–civil-society divide. It is this partial overlapping of institutions, both in terms of membership interests and function, that characterizes the institutional tissue of such areas as 'thick'. This dense mutual interpenetration of key individuals, interests, and institutions is crucial in shaping views of what is desirable and possible by way of stasis and change in such areas and so in reproducing those same institutional structures, especially in conditions of economic well-being.

3. In some cases, such as the United Kingdom Enterprise Zone experiments, new central government small-area policies were specifically created with the intention of promoting localized solutions, buttressed by central government expenditures, while presented as evidence of an emerging enterprise culture.

4. Certainly the evidence from areas of successful small-firm growth elsewhere in Europe would support such a conclusion (see e.g. Hudson and Williams 1989).

5. The London Docklands UDC operates on a different scale from the remaining nine. In 1990–1, for example, its budget from government was £333 m, while that of the remaining nine was £550 m. Moreover, its proximity to the City of London offers a very different environment from those of the other UDCs.

6. In practice, the distinction between state and market is at best one of degree; markets are social constructions, which depend heavily on state activity and regulation, as the events of the 1980s in the United Kingdom and elsewhere sharply revealed.

REFERENCES

Austrin, T., and Beynon, H. (1979), 'Global Outpost: The Working Class Experience of Big Business in North East England, 1964–1979' (Dept. of Sociology, University of Durham).

Beynon, H. (1985), *Digging Deeper* (London: Verso).

Byrne, D. (1990), *Beyond the Inner City* (Milton Keynes: Open University Press).

Caulkins, D. (1992), 'The unexpected entrepreneurs: Small high technology firms and regional development in Wales and North East England', in F. A. Rothstein and M. Blim (eds.), *Anthropology and the Global Factory* (New York: Bergin and Garvey).

Dicken, P. (1990), 'Seducing foreign investors: The competitive bidding strategies of local and regional agencies in the United Kingdom', in M. Hebbert and J. C. Hansen (eds.), *Unfamiliar Territory* (Aldershot: Avebury).

—— (1992), *Global Shift* (London: Paul Chapman).

DIDA and DDC (n.d.), *Derwentside Fact File* (DIDA and DDC, Consett, County Durham).

Douglass, D., and Krieger, J. (1983), *A Miner's Life* (Andover: Routledge and Kegan Paul).

Firn, J. R. (1975), 'External control and regional development: The case of Scotland', *Environment and Planning A*, 7: 393–414.

Grabher, G. (1990), 'On the weakness of strong ties: The ambivalent role of inter-firm relations in the decline and reorganisation of the Ruhr' (WZB Discussion Paper FS I 90–4; Berlin).

Hassink, R. (1992), *Regional Innovation Policy: Case Studies from the Ruhr Area, Baden Württemberg and the North East of England* (The Netherlands Geographical Studies, 145: The Hague).

Hewison, R. (1987), *The Heritage Industry* (London: Methuen).

Hudson, R. (1989), *Wrecking a Region* (London: Pion).

—— (1991), 'The North in the 1980s: New times in the "Great North" or just more of the same', *Area*, 23/1: 47–56.

—— and Sadler, D. (1983), 'Region, class and the politics of steel closures in the European Community', *Society and Space*, 1: 405–28.

—— —— (1986), 'Place, class and the uneven development of capitalist societies', in M. Storper and A. J. Scott (eds.), *Production, Territory, Work* (London: Allen and Unwin).

—— —— (1989), *The International Steel Industry: Restructuring, State Policies and Localities* (London: Routledge).

—— —— (1991), 'Manufacturing success? Reindustrialization policies in Derwentside in the 1980s' (Occasional Publication 25; Dept. of Geography, University of Durham).

—— and Townsend, A. (1992), 'Trends in tourism employment and resulting policy choices for local government', in P. Johnson and B. Thomas (eds.), *Perspectives on Tourism Policy* (London: Mansell).

—— and Williams, A. (1989), *Divided Britain* (London: Belhaven).

—— Krogsgaard Hansen, L., and Schech, S. (1992), 'Jobs for the Girls? The New Service Sector Economy of Derwentside in the 1980s' (Occasional Publication 26; Dept. of Geography, University of Durham).

Judd, D., and Parkinson, M. (1990) (eds.), *Leadership and Urban Regeneration* (London: Sage).

Lash, D., and Urry, J. (1987), *The End of Organised Capitalism* (Cambridge: Polity).

MacDonald, R. (1991), 'Runners, fallers and plodders: Youth and the enterprise culture', Paper presented to the 14th National Small Firms and Policy Conference, Blackpool.

Mair, A. (1991), 'Just-in-time manufacturing and the spatial structure of the automobile industry', *Change in the Automobile Industry: An International Comparison* (Discussion Paper; Dept. of Geography, University of Durham).

Mandel, E. (1978), *The Second Slump* (London: New Left Books).

Martin, R. (1988), 'Industrial capitalism in transition: The contemporary reorganisation of the British space-economy', in D. Massey and J. Allen (eds.), *Uneven Re-development: Cities and Regions in Transition* (London: Hodder and Stoughton).

Massey, D., and McDowell, L. (1984), 'A Woman's Place?', in D. Massey and J. Allen (eds.), *Geography Matters!* (Cambridge: Cambridge University Press).

Offe, C. (1975), 'The theory of the capitalist state and the problem of policy formation', in L. N. Lindberg, R. Alford, C. Crouch, and C. Offe (eds.), *Stress and Contradiction in Modern Capitalism* (Lexington, Ky.: DC Heath).

Parkes, C. (1991), 'Dying heart still beats', *Financial Times* (6 Dec.).

Pugliese, E. (1985), 'Farm workers in Italy: Agricultural working class, landless peasants or clients of the welfare state?', in R. Hudson and J. Lewis (eds.), *Uneven Development in Southern Europe* (Andover: Methuen).

Rees, G., and Thomas, M. (1989), 'From coalminers to entrepreneurs? A case study in the sociology of reindustrialisation', Paper presented to the British Sociological Association Annual Conference, Plymouth.

Sadler, D. (1992), *Global Region* (Oxford: Pergamon).

Sengenberger, W., and Loveman, G. (1987), *Small Units of Employment* (Geneva: International Institute for Labour Organisation).

Spellman, J. (1991), 'Attracting Inward Investment: The Strategies of a Regional Development Organisation', MA thesis (University of Manchester).

Storey, D. (1990), 'Evaluation of policies and measures to create local employment', *Urban Studies*, 27: 669–84.

Thompson, E. P. (1968), *The Making of the English Working Class* (Harmondsworth: Penguin).

10

Local and Regional Broadcasting in the New Media Order

Kevin Robins and James Cornford

INTRODUCTION

To consider the media is to simultaneously raise questions of economics—production, distribution, and consumption—and questions of culture—meanings, identities, and ways of life. In the present period, both the media industries and media cultures are undergoing complex, and sometimes contradictory, processes of globalization. In this chapter, we trace out some of the consequences and implications of these processes for localities and regions. Using the example of place-based media policies and strategies in Britain, we are concerned to show how places are being forced to reconfigure and reimagine themselves in the context of the emerging global media order. In our corner of the world, a key term in these attempts has been 'Europe': Europe as a successor to the national space in terms of the opportunities it offers both for a new cosmopolitanism and for a new regionalism. Europe offers the ambiguous possibility for regions to be both 'in and against' the global order. What these strategies fail to come to terms with, as our case studies show, is the reality of Time Warner, Sony, or CLT which exposes the weaknesses and idealism of any such aspiration. In so far as the activities of these global behemoths are coming to shape and reshape the possibilities for developing local media industries and services, they also, and perhaps more importantly, reshape our conception of what locality is. As Joshua Meyrowitz (1989: 332–3) argues, in today's media order 'experience is unified beyond localities and fragmented within them' with the result that 'we are no longer "in" places in quite the same way' as we once were.

GLOBALIZATION AND THE NEW MEDIA ORDER

It is strange that so little attention has been given to the geography of the mass media. To take the case of broadcasting and audiovisual media, it is clear that programmes are produced in particular places, and then distributed—via broadcasting, satellite, microwave, cable, cassettes—across space or spaces. At various times, in various contexts, the issue has been implicit: in international debates about media imperialism or programme flows, for example; or, more locally, in the sporadic initiatives that have developed around community media. What never emerged was a more systematic account of the significance of 'audiovisual spaces', of the relation between television and territory. There are signs, however, that this state of affairs might be changing. Contemporary developments, we shall argue, have made it clear that questions of geography are increasingly central to any adequate understanding of media economics and politics (Robins 1993).

In Europe, at least, we might presume that it was the absolute prevalence of national broadcasting systems that made us insensitive to broadcasting geographies: the apparently universal, and even natural, quality of the national audiovisual space obscured or disavowed its constructedness and thereby the arbitrariness of its 'nature'. It is perhaps difficult for us to recall that the tradition of public-service broadcasting, which prevailed for over sixty years in Europe, was not 'in its nature' a national, or nationalist, phenomenon. Even in the earliest days, the technologies and economics of broadcasting pushed towards international-ization. As Stig Hjarvard (1991: 4) has argued, 'television has from its very start been an international medium. Neither in relation to technology, institutional structure nor conventions of expression does it make very much sense to call radio and television national media.' Broadcasting on a scale below that of the national was also a significant issue and more than a possibility in those early days; in Britain, as in other European countries, broadcasting actually began on a local basis from a diverse range of stations in provincial cities, and localism and regionalism were issues of keen debate (see e.g. Scannell and Cardiff 1991: chs. 14, 15; Pegg 1983: ch. 6; Briggs 1975).

In the development of the public-service tradition, however, such possibilities were decisively rejected. The national cultural space was insulated as much as possible from what were seen as the insidious forces of commercialism—symbolized by Radio Luxembourg and by 'vulgar' American programming—which seemed to threaten the 'integrity' of the British way of life. And, within the national culture, the values and attitudes of regionalism and localism were subordinated to a policy of centralization: 'There was a retreat away from direct contact with listeners, from their participation in programmes, from informality, friendliness and easy accessibility into a distanced, anonymous collective voice' (Scannell and Cardiff 1991: 317). The clear objective was to create a culture of national unity, and 'though something like a common national culture and identity was given expression in moments of ritual celebration, it was often at the expense of different cultures and identities within the imposed unity of the United Kingdom and its national broadcasting service' (ibid. 303). It was largely through the determination of the first Director General of the

BBC, John Reith, that the broadcasting system was centralized and nationalized. His great achievement was to create a system that was in all respects 'British': a nationalized industry, functioning as a national public sphere, and articulating a particular conception of national culture and identity. That this came to seem for so long, and not just in Britain, the natural and given form of existence for broadcasting is testimony to what was accomplished.

But times are changing, and what once seemed 'natural' no longer seems so. Over the past decade or so, the values and principles of the public-service tradition have become subordinated to those of the market. The complex interplay of economic, technological, and regulatory change during the eighties has brought about dramatic transformations in audiovisual industries and cultures and laid the basis of what must be seen as a new media order. In this new order, there has been an important renegotiation of political and economic roles: as Pierre Musso and Guy Pineau (1989) have argued, we have seen a decisive shift from state regulation, with its more politically oriented objectives, to a new 'entrepreneurial leadership' of the media industries which is driven primarily by economic and business imperatives. Through this process, the social and political concerns of the public-service era—with democracy and public life, with national culture and identity—have come to be seen as factors inhibiting the development of new media markets. The new media order is about the rise to power of private-sector vested interests, and the overriding priority of those interests has been to dismantle such 'barriers to trade'.

It is here that new questions of geography become central. As Geoffrey Nowell-Smith (1991: 42) remarks, in the old public-service model 'the audience is defined as the nation, bounded territorially by national frontiers and then subdivided regionally or by other recognized forms of difference'. It was this that seemed the 'natural' geography of broadcasting, but, as we are now coming to recognize, it represents only one possible arrangement. As Nowell-Smith points out, 'it is possible to conceive of broadcasting in a totally different way, one which considers audiences economically rather than politically and subdivides them according to their accessibility as parts of a consumer market whose frontiers are by no means necessarily those of nation-states'. Broadcasting has been fundamental to the production of national territory, and to the construction of the unified identity of national community and tradition; now, however, as is becoming increasingly apparent, it is implicated in the creation of other territories, other kinds of community, other ways of belonging. Audiovisual geographies are becoming detached from the symbolic spaces of national cultures and realigned on the basis of the more 'universal' principles of consumer demographics and market segments.

The imperative has been to break down the old boundaries and frontiers of national cultures, which have come to be seen as arbitrary and irrational obstacles to the expansion of media markets. The idea of television without frontiers, the free and unimpeded circulation of programmes, has become the guiding principle of the new media order. It is an idea whose logic pushes towards the ideal of creating global markets and programming, and we are already well aware of global corporations—the likes of Time Warner, Sony, and News Corporation—who are intent on turning this ideal into a solid reality. 'The new reality', in the words of Theodore Levitt (1983: 21), 'is the globalization of markets, and with it the powerful materialization of the global corporation.' The new media order is a global order.

Such a statement requires qualification, of course. The first thing that should be emphasized is that a global media culture is far from being established and the old order is still a long way from being defunct. The evidence suggests that television still remains very much a national medium and that national cultures of television will continue to be a force to be reckoned with (Sepstrup 1989; Bourdon 1992). As one French observer remarks, 'European publics want national products'; 'more profoundly', he adds, 'culture, and particularly audiovisual culture, touches the very heart of nationalism' (Cluzel 1992: 46). Global media still have a long way to go in negotiating the problem of national differences and particularisms. They also have to come to terms with legal and political barriers erected by governments and other regulatory agencies to protect national interests. In Europe, there are quotas imposed on non-European (that is, in effect, American) programming, which are aimed at both protecting cultural sovereignty and enhancing the competitiveness of domestic producers. In the United States, where the forces of globalization have been experienced through the symbolic take-over of Hollywood studios by Sony and Matsushita, reaction takes the form of growing concern about rules of foreign ownership. If globalization is the prevailing force at work in the media industries, then there are significant forces of resistance that preclude its absolute prevalence.

There is also a second, and perhaps more important, qualification that should be made to the globalization thesis: the new media order is a global order, but it is not only, or rather it is not straightforwardly, a global order. We must recognize that the globalization of markets has contradictory implications for culture and identity. The more social life is subjected to the forces of global consumerism, Stuart Hall (1992: 303) argues, 'the more identities become detached—disembedded—from specific times, places, histories, and traditions, and appear "free-floating" '. This may well be experienced in terms of release from the closure of old national identities, and there may be a sense of new and cosmopolitan possibilities being opened up by the globalization of culture. Identities become 'more positional, more political, more plural and diverse; less fixed, unified or trans-historical' (ibid. 309). But there may well be a darker side to the logic of globalization. Zygmunt Bauman (1992: 692–3) draws attention to the same phenomenon as Hall when he observes that contemporary life is characterized by its condition of 'territorial uprootedness', and that identity-formation has been 'de-territorialized'. But, as Bauman suggests, this is equally likely to provoke anxiety and the search for an alternative kind of rooted identity. The problematization of national belonging 'does not mean that the identity-constructing function in which the nation states used to specialise, is likely to fade off together with its carriers. [It] only means that the function will probably seek another carrier' (ibid. 692). One easily available carrier is what Bauman describes as 'the territory classified as *culture*' and as 'the contrived, made-up community masquerading as a Tönnies-style inherited *Gemeinschaft*' (ibid. 696). Identities that feel they are becoming detached and disembedded through the forces of globalization may readily seek refuge in new kinds of communal coherence and cohesion.

EUROPE IN THE NEW MEDIA ORDER: MELTING-POT OR MOSAIC?

It is because the initial costs of production represent the principal investment, and because the subsequent costs of reproduction or transmission for larger audiences are comparatively small, that there has always been an essentially expansionist logic at work in the media commodity and media business. And because audiences are geographically distributed, this has, of course, always had geographical implications; there has always been a push towards the creation of enlarged audiovisual spaces. For several decades, as we have already argued, this expansionist dynamic was resisted in Western Europe by governments that prioritized the political and cultural agenda of national public-service broadcasting systems. Now, it seems, these powers of resistance are considerably weakened. In the face of an economic and technological logic that shows no respect for the formalities of national frontiers and boundaries, it has become necessary to find some new accommodation for political and cultural interests.

What we have seen over the last ten years is the fulfilment of the logic of expansion by organizations whose aim is to compete in world markets and whose priority is to achieve the scale economies that make this possible (Aksoy and Robins 1992). The world's new media order is being constructed through the entrepreneurial devices of global corporations and conglomerates—the likes of Time Warner, Sony, Matsushita, News Corporation, Philips, Fininvest, CLT, Bertelsmann, and Walt Disney. Steven Ross, the late head of the massive Time Warner, put it all quite clearly in his 'Worldview Address' delivered at the 1990 Edinburg International Television Festival. 'The new reality of international media', he told his audience, 'is driven more by market opportunity than by national identity.' Time Warner stood for 'the free flow of ideas, products and technologies', he said, and 'that mission will recognize no territorial borders—just as creativity knows no borders'. In our industry, Ross went on, 'the mergers and partnerships that have recently begun to reshape the map will continue—in fact, I predict they will accelerate—and they will do so regardless of borders' (Ross 1990: 12–16). According to the US Department of Commerce's National Telecommunications and Information Administration, 'public opinion on a worldwide basis is increasingly shaped by global media' and this trend towards globalization may be seen

as leading inevitably to a new world order in which the significance of political borders, national identities and regional and cultural differences is diminished through information and entertainment distributed by global firms. According to this scenario, experiences shared on a global scale through communications media will eventually transcend differences among citizens of separate nations or regions within a nation (Obuchowski 1990: 7, 45).

This chimes precisely with the Time Warner world-view. And it is also a view shared by one of the most vocal critics of transnational media:

How aptly expressed are the current objectives of the for-the-moment unrestrained global corporate order—open borders, which can be transgressed; open trade, which enables the most powerful to prevail; open minds, which are at the mercy of the swelling global flows of the cultural industries. At least for now, the celebratory mood seems justified (Schiller 1991: 16).

There is, then, a growing sense that the future is about global markets, and that global markets are inevitably about the dissolution of old cultures and identities.

It is this future that broadcasting policy in Europe is now having to come to terms with. The challenge has been taken up with a great sense of urgency. How to position European interests in the new media order? What we are seeing is the emergence of a new supranational regulatory environment in which the emphasis has shifted dramatically towards questions of economic, industrial, and competition policy, and away from the political and civic concerns that characterized the old public-service systems. Now the emphasis is on the creation of the large European audiovisual market that will eliminate barriers to the buying and selling of programmes and to their transmission and reception within the European Community. 'National film and television industries in Europe must change,' one European Commission policy-maker declares; 'they will have to adopt transnational forms of production, financing and distribution which will necessarily alter their present nature.' In the new order it will be imperative to achieve 'critical size': 'it seems inevitable that the European audiovisual scene will be dominated in the not too distant future by a few giants controlling the whole spectrum of the sector segments, from production to distribution. Vertical concentration will represent the actual formation of an audiovisual sector in the first place' (Maggiore 1990: 39, 79).

New media technologies and markets make a mockery of borders and frontiers; in Europe, too, the order of the day is now the 'free circulation' of audiovisual products and services. There is the conviction that Europe must move rapidly from its old condition of national fragmentation to a new state of integration and cohesion, and it is the likes of CLT and Berlusconi that promise to make this possible. 'The depth of change', we are told, 'its speed and the stakes involved should dissolve trivial preoccupations for the preservation of autonomy and induce an urge for common, genuinely common, action' (Maggiore 1990: 131). What is so urgent is to make the transition from a model of regulation which required broadcasters to provide a diverse and balanced range of programmes (education, information, and entertainment) for citizen-viewers, to a successor model in which the imperative is to maximize the competitive position of European media businesses aiming to satisfy the needs of consumers in global markets.

National barriers are now seen as 'a major handicap for Europe's industry and cultural identity' (CEC 1986: 6). But what are the real implications of the proposals for a new regime of 'television without frontiers'? What happens to broadcasting economies and cultures when national markets and communities begin to be dismantled? Some twenty years ago now, in a discussion of the European film industry, Thomas Guback (1974: 10) identified what he understood to be a dominant logic at work in the European project. The creation of an economically integrated Europe, he maintained, 'favours the enlargement of firms to international stature, with concomitant trends toward standardisation, at the expense of small enterprises and a great deal of variety'. And if this is the case, he went on,

then it is obvious that the major emphasis is not upon *preserving* a variety of cultural heritages, but rather upon drawing up a new one which will be in tune with supranational economic considerations. In that case, we had better forget about the past and concentrate upon seeing the creation—or fabrication—of a new economic European consumer whose needs will be catered to—if not formed—by international companies probably operating with American management and advertising techniques (ibid. 10–11).

In Guback's view, 'changes in the economic and political realms . . . necessarily have an impact on the cultural sphere, tending to transform a mosaic into a melting pot' (ibid. 10). This, it seems to us, is still a compelling logic, still powerfully at work, in the European media industries. The expansion of markets seems to work—or is seen as working—inevitably towards the detachment of media cultures from particular places, histories, and traditions.

If this expansionist economic logic is the dominant force at work in the European media industries, there is also a countervailing force that must be taken into account. After documenting the trends towards standardization and homogenization in these industries, Guback then goes on to argue that 'the aim must be to preserve the mosaic of cultures and to resist the temptation to rely upon bigness itself as a solution—even in the face of seemingly overwhelming political and economic trends' (1974: 16). In the light of accelerating tendencies towards economic and cultural integration, this argument, or exhortation, has come to have a growing resonance. In the face of those forces that have promoted centralization and homogenization, there are growing appeals to the importance of community and identity in the new Europe. And it is to the ideal of regionalism—to the diversity and difference of a 'Europe of the regions'—that these appeals are being made.

Against the principle of the melting-pot, the advocates of European regionalism are fighting to sustain the image of the continent as a cultural mosaic. It is in this context that local and regional media are coming to be seen as fundamental resources of both community and identity. A recent Council of Europe report, for example, makes a stark contrast: on the one hand there is the 'centralised, multinational, standardised "mega-communication" ' of national and international media, with its 'commercial logic'; and, on the other, regional ' "meso-communication", which is decentralised, disinterested, serves the community and involves cultural and pluralist diversification' (Musso 1990: 4). Another observer draws attention to the fact that growing internationalization has been shadowed by a renewed interest in regional broadcasting. 'A new idea is coming to light', he suggests, 'that regionalism can be an antidote to the internationalization of programming; that it will prove indispensable in compensating for the standardisation and loss of identity of the big national [and international] networks' (Trelluyer 1990: 11). The appeal is clearly to the kind of situated meaning and emotional belonging that is felt to have been eroded by the forces of internationalization and then globalization.

In the evolution of European Community broadcasting policy both integrationist and regionalist strategies have figured—often alternating in the role of the dominant discourse. In the early and mid-eighties, particularly, what was being put forward was a strategy to create the large audiovisual market that—it was believed—would sustain an equally large audiovisual industry, competitive on a global scale with American and Japanese interests. This, it was assumed, would be predicated on the development of a European common culture and shared identity. What this in fact represented was an expanded version of the national broadcasting model, one that sought to maintain, at a higher level, the congruence between economic, political, and cultural spaces of broadcasting. The expectation was that a European audience might come to enjoy the imagined community and solidarity of some kind of supranational political and cultural identification. As such, it expressed the desire to achieve coherence and cohesion for the pan-European idea. By the late eighties, however, the mood had changed considerably and this unitary model of Europe had lost momentum. There has been an increasing emphasis on diversity over unity. The European Commission's MEDIA programme, which provides loans and support for small producers and distributors across Europe, represents the most concrete expression of this emphasis. In this case, policy and strategy have been concerned with recognition of, and sensitivity towards, cultural difference. 'We have no interest in promoting a melting pot', says one EC policy-maker, 'we want to preserve European identities' (quoted in Collins 1992: 14). Growing anxiety about the 'dilution' of culture as a consequence of integration and unification has meant that there is now a renewed interest in the particularity and variety of regional and national cultures in Europe (Robins 1989; Morley and Robins 1989).

This tension between unity and diversity—expressed in the images of melting-pot and mosaic—must be central to any account of the transformations now going on in Europe. What is at issue is the relationship between openness and closure of economies and cultures. The historian J. G. A. Pocock (1991: 9) sees European integration in terms of the omnipotence of market forces and of the 'surrender by states of their power to control the movement of economic forces which exercise the ultimate authority in human affairs'. What we are seeing is the creation of 'administrative or entrepreneurial institutions designed to ensure that no sovereign authority can interfere with the omnipotence of a market exercising "sovereignty" in a metaphorical because non-political sense'. It is in response to the growing hegemony of transnational and global markets and the (partial) dismantling of national state mechanisms that the question of 'sovereignty' is now being raised at the level of localities and regions in Europe. Reluctant to align their interests with those of global corporations, cities and regions are making claims to some degree of control over their own destinies. Against the logic of 'de-territorialization'—to use Bauman's terminology—we are seeing the development of initiatives aimed at the 're-territorialization' of economies and cultures. Media industries and cultures exemplify both the vigour and the vicissitudes of such initiatives, and it is to a discussion of the limits and possibilities of local and regional broadcasting strategies that we turn in the following sections.

THE LOCAL IN THE GLOBAL: MEDIA DEVELOPMENT
STRATEGIES IN BRITAIN

In Britain during the 1980s, this agenda for the 're-territorialization' of broadcasting became an important focus of attention, with renewed interest in local and regional media in terms of both their economic and cultural significance. At a conference on 'Cities and City Cultures', held in Birmingham in 1987, one participant drew attention to the cultural industries agenda. There was, he said,

a renewed interest in cultural development at the local and regional level, principally generated by the economic recession, the decline in manufacturing industries and investment in new growth sectors, particularly the service sector. Within this framework cultural development is drawn into relationship with a number of interests whose common concern is regeneration of local economies. In effect, policies for cultural development based on cultural industries are beginning to emerge as part of a wider logic of economic renewal (Hurd 1987: 3).

And, as another contributor to the same conference made clear, such an approach was in defiance of the logic of global markets: 'Maxwell, Murdoch, and others are not going to wait for us. They see money to be made in these industries. The free market will overtake us unless we change our attitudes and plan strategies very quickly indeed. If we do not grasp this opportunity it will be wrested from us' (Fisher 1987: 2). What seemed to be emerging was a new context for the development of local and regional broadcasting, one in which local development agencies and programme producers would become allied in the cause of promoting both economic and cultural agendas.

The immediate catalyst for local media development in Britain was the creation of Channel 4 in 1982. Unlike the existing broadcasters, Channel 4 made no programmes itself, but rather commissioned much of its schedule from a new sector of 'independent' television production companies which sprang up to service it. This new development in British broadcasting was driven by two forces: a cultural agenda concerned with diversity and innovation; and an economic agenda driven by a neoliberal (Thatcherite) desire to promote competition, flexibility, and efficiency and to break up the vertically integrated national broadcasting systems (BBC and ITV). It was this new sector of small firms—the independent producers and associated service companies, oriented around Channel 4—that, for a short time at least, gave the impression of a countervailing principle to that of concentration and globalization in the media industries. Most importantly, although independent production was strongly concentrated in the capital, there were virtually no provincial cities which did not see the emergence of a few of these small television production companies, often with strong aspirations to make programmes focused around their locality. As such, these companies appeared to local economic development interests as both deserving of, and requiring, public support. In short, they seemed to offer certain possibilities for local interests to intervene in this broadcasting agenda for their own ends (Robins and Cornford 1992).

The earliest and most influential attempt to develop a local media strategy was undertaken in London in the early 1980s by the Greater London Council. While the GLC was acutely aware that the audiovisual industries were dominated by large multinational companies which maintained their position through oligopolistic control of the distribution of cultural goods and services, it insisted that 'the alternative of flexible specialization is beginning to make an impact in the cultural industries' (GLC 1985: 179). There was a will to see, in 'the growth of a new generation of small independent producers', signs of 'the faltering emergence of a regime of "flexible specialisation" '. It was an approach that brought together a constellation of ideas—post-industrialism, local entrepreneurialism, the increasing importance of cultural facilities for the development and regeneration of places, flexible specialization, industrial districts—to constitute a new approach to urban and regional policy based on the idea of local development through the fostering of indigenous small firms (Murray 1991).

This approach, derived from the experience and conditions of London, was readily taken up in many of Britain's provincial cities and regions, leading to the development of local media development strategies—notably in Manchester, Birmingham, Liverpool, Sheffield, Newcastle, Edinburgh, and Cardiff—often as part of wider media or arts and cultural industries sector interventions. Whether explicitly or implicitly, all these media development strategies were influenced by ideas of flexible specialization (which, as a 'theory' of economic change, was often used to give a certain credibility and legitimacy to what were generally unusual and unorthodox ideas in local government circles).

By the mid-1980s this was giving rise to two particular lines of action in cultural industries policy, both derived from the tenets of the flexible specialization model. The first of these was the (micro-corporatist) idea of local media development agencies. The aim of such agencies was to manage the collective or public provision of common services (e.g. training, distribution) and infrastructure (e.g. post-production facilities), and also to ensure the promotion of networking through the production of directories and the creation of some kind of local forum for the industry to share information. Birmingham, for example, established a Media Development Agency with the backing of public and private sectors, and, in Cardiff, South Glamorgan Council backed the creation of a development agency, 'Cardiff Media City'. In some areas, the focus was more on the provision of the physical and technological infrastructure of studios and editing suites. Sheffield, for example, established an Audiovisual Enterprise

Centre offering a range of services to the local independent production sector. In other areas, where the basic technical infrastructure was available, initiatives sought to provide collective services such as consultancy, training, or distribution. Thus, in Newcastle, the North East Media Development Trust established the North East Media Development Agency with a brief to provide consultancy and advice. The second line of action was the planning of quarters, zones, and centres to stimulate the clustering of audiovisual and other cultural industries companies. This idea of cultural quarters designed to promote 'flexible' interaction between cultural producers helped to build a bridge between local economic development and urban regeneration strategies. In Sheffield, for example, a Cultural Industries Quarter was established near the city centre, and in Birmingham the Digbeth area was designated a Media Zone.

If these local initiatives were understood as the kind of interventions necessary for promoting the media industries in the regions, the late 1980s saw a very significant shift in the media development agenda. The focus began to move away from an almost exclusive attention to the links between indigenous small firms within the locality or region, and towards a concern with links between local firms and the national and international centres of power within the audiovisual industries. This, we would argue, marked a recognition of the increasingly globalized nature of the media industries (and of the accelerating pace of liberalization which was opening up British broadcasting to these global forces). From the late 1980s there has been an implicit, and often explicit, acknowledgement that local industries must be 'plugged in' to global distribution systems (national and international broadcasters and the Hollywood film studios).

The most prominent manifestation of this new orientation to the national and international sphere has been the proliferation of screen commissions and location bureaux, strongly influenced by US and Canadian models (Boden 1990; 1991). The main role of these organizations is to market a city or region as a desirable location for audiovisual production, attracting film and television productions into the area. Feature film and big-budget television drama productions shot in the area, it is argued, generate significant expenditure in the region, both in terms of the audiovisual sector (camera operators, lighting technicians, set construction, extras, editing studios, laboratories) and more generally in the service sector (hotels, transport, catering). In order to induce 'outside' producers to use the area as a location, the screen commission undertakes to ease the relations between producers and the locality, for example by finding appropriate locations (e.g. a castle), seeking permissions (e.g. to close off a road), and generally acting as intermediary. A related development has been the emergence of regional film investment funds, influenced, in particular, by German models (Boden 1988). Such funds offer a financial inducement to use particular locations by offering, on commercial terms, to make a given investment in a project in return for a guaranteed proportion of the budget being spent in the area. Screen commissions and investment funds are, then, largely concerned with attracting inward expenditure to the region.

A number of examples of this new type of strategy can be given. There has been the establishment of Scottish Screen Locations in 1990 as the first step towards developing a Scottish Screen Commission. At a smaller geographical scale, the local-authority-financed Lothian Screen Industries Office in Edinburgh aims to attract productions to the region. In England, the Northern Media Forum has sought the backing of local authorities in establishing both a Northern Screen Commission (successfully) and a Northern Film Investment Fund (unsuccessfully). In Liverpool, the idea of attracting productions from elsewhere has been pursued by the appointment of a local-authority-financed Film Officer. A recognition of the importance of this kind of approach has come with the decision by the Department of Trade and Industry to put £3.5 m into funding the British Film Commission (Brooks 1993). There have been attempts to regionalize some of the activities of this new body, but how far this process should go has been contested by those who argue that the priority is to achieve critical mass for a national audiovisual industry centred in London.

There has been, then, a general reorientation away from the early 1980s strategies, based on the promotion and local networking of small independent production firms, towards a new agenda that seeks to promote links with large firms and markets outside the region. This reflects an increased awareness of the strategic importance of distribution and finance in the audiovisual industries. These developments should be seen as representing a more 'realistic' accommodation to an increasingly concentrated and globalized media industry (Cornford and Robins 1992).

THE GLOBAL IN THE LOCAL? THE STORY OF CHANNEL 5

All this was the broader context in which attempts have been made over the last three or four years to legislate Channel 5 into existence. As the last terrestrial channel in Britain, and therefore almost certainly the last channel wholly within the British regulatory sphere, Channel 5 has been held up as offering the possibility of introducing something really new into British broadcasting. Could this new channel, like Channel 4 before it, extend and enhance the ideals of public-service provision? This was the question that was asked of it.

Two affirmative proposals emerged in response. Each brought a geographical perspective to the broadcasting agenda; and, more particularly, each, in its different way, saw the development of cities and regions as the new focus for public-service objectives. The most pragmatic approach came from what was effectively an anti-metropolitan lobby, arguing that Channel 5 could be (almost) anything it chose to be, provided it was based outside, and preferably a long way outside, London. Most of the activity from this lobby was driven by local authorities and develop-ment corporations, and by the various media development agencies that had emerged in the mid-1980s in Britain's major provincial cities (those in Sheffield, Edinburgh, Glasgow, and Liverpool being most vocal). Televi-sion was seen as a small, but modern and expanding, industry, and Channel 5 was seen as fitting into the rationale of local or regional economic development as a major inward investment opportunity. The channel was expected to provide benefits in terms of bringing jobs and skills to the chosen location, and there was the anticipation that it would stimulate local programme producers and generate the critical mass of activity necessary to sustain a viable media industry.

There were also cultural objectives. Channel 5 was seen as a chance to develop one or more regional centres producing programmes for national and international markets. Such centres, it was argued, could act as a counterweight to the concentration of television production, and therefore of cultural and political influence, in and around London. Though it could not legislate on the location of the new channel, the Independent Broadcasting Authority, the regulatory body responsible at that time for Channel 5, itself recognized that there would be 'some advantage' if its headquarters were in a major regional city since this would 'encourage a distinctive non-metropolitan identity for Channel 5, as well as local sources of programme supply' (IBA 1989).

In areas such as Newcastle and Glasgow which already had a small television industry, the new channel was seen as an opportunity to expand the industrial base and to develop critical mass. For places which had no ITV company or major BBC studios, such as Sheffield and Edinburgh, it was largely seen as a way of getting on to the media map. What was common to all these places, however, was a determination to overcome what they saw as the deep historical imbalance which had concentrated resources, decision-making, and employment in London. For the more ambitious cities, the aim was to position themselves as major players, not only in the British, but also in the European, media landscape.

An alternative, and far more idealistic, model for the organization of Channel 5—and one with more of an emphasis on service provision than economic development—was a proposal to use the new channel as a vehicle for introducing a local television service. At the Royal Television Society convention in September 1989, Chris Rowley (the Head of Television Planning at the IBA) outlined the possibility of part-local, part-national service with thirty to thirty-five local stations 'opting-out' of the national service for a few hours a day in order to broadcast local programmes. His calculation was that a local service could be sustained in towns and cities with a minimum of 300,000 population per station (Rowley 1990). Rowley called this model for the new channel 'city-share television', and he went on to show his faith in its viability when he resigned from the IBA and set up a consortium, Five TV Limited, to bid for the new channel franchise.

This local–national model for Channel 5 rapidly gained an impressive degree of support. Those who wanted to maintain some degree of continuity with the tradition of public-service broadcasting seized upon Rowley's proposal as the best way to prevent the channel from becoming a predatory commercial enterprise. Regionalism and localism were clearly seen as holding out the best prospects for averting this outcome. It was these advocates of a local option for Channel 5 who were most successful in shaping the policy debate over the new channel. There was an almost instictive sense that 'local' must be a good thing; and, of course, the 'city-state' label evoked images of local democracy and civic identity. Indeed, so appealing was the ideal, that the debate increasingly centred on whether the city-state model was economically viable, and whether there was sufficient potential in local advertising revenue to sustain the 'opt-out' stations. What was not clearly addressed, however, was why such a model was desirable, and whether this really was the most appropriate way to develop public-service objectives through the new channel. What is 'local' in the new media order? This is the question that was never seriously addressed (Robins and Cornford 1991).

These, then, were the initial ideas being proposed for a 'distinctive' Channel 5. But what happened when the bidding process came into play and the new channel was finally and really on offer? What would the key players in the industry, the media financiers and investors, think of the prospects for a channel that would have to survive in direct competition not only with the other terrestrial channels (ITV, Channel 4, and the BBC), but also with Rupert Murdoch's BSkyB satellite channels?

What we saw was the sudden arrival on the scene of a range of interests with no known interest in regional development issues or in local television services. Suddenly, the debate about how Channel 5 might contribute to public-service agendas was rudely interrupted by the commercial and competitive activities of global media concerns—none other than Time Warner, Sony, and Silvio Berlusconi's Fininvest. Berlusconi was the first to make a move when he joined a long-established consortium that had originally intended to locate the channel's headquarters in Glasgow. Following this, a newly created consortium, the Entertainment Channel, came to the fore with backing from the Daily Telegraph Group and also from the mighty Time Warner. And, all the while, Chris Rowley's Five TV was still in the race, and it, too, was now starting to keep company with the big league, joining forces with Sony, through its subsidiary Columbia Pictures, with Canadian broadcaster Canwest, and with Thames Television.

Channel 5 had, then, rapidly become a matter of interest to the most powerful global players (Masters 1992).

It was not to last, however. As they surveyed the British scene and assessed the prospects for the new channel, these various organizations all began to retreat. Berlusconi was first to go, bringing about the collapse of the consortium with which he was associated. Then, as the Entertainment Channel's business plan began to look more and more precarious, Time Warner withdrew its involvement. And before long, Sony and Canwest, too, had decided that their money could be better invested elsewhere. So what had immediately become clear was that Channel 5's success would depend on the backing of global players in the media business. And what became clear more gradually was that those players did not rate the prospects for the channel's success very highly. What chance then for all the smaller interests who had invested such great hopes in what Channel 5 might mean for the cause of Britain's cities and regions?

And then a new twist to the story. When the bids were finally submitted to the Independent Television Commission (the successor to the IBA as regulator) in July 1992, only one resilient contender was still left in the race—Five TV. Now renamed Channel Five Holdings, it had joined forces with Thames Television and, more significantly, with Moses Znaimer's Citytv, Toronto. In its proposals for the channel (see Channel Five Holdings 1992), what was on offer was a new model of what local television could be, a model very much based on the North American example. It is, in fact, better described as global-local television. What was promised was city television initially for the capital, CityTV London, with the possibility of future developments in Manchester, Birmingham, and Glasgow or Edinburgh. Programming would have a global-local flavour, with a diet of local (meaning London) news combined with the global fare of movies and music television. Thames Television's building was to be extensively rebuilt, modelled on Citytv's Toronto headquarters.

There were reports that Fininvest was back on the British scene, with Berlusconi coming close to signing up as a major shareholder in Channel Five Holdings. But then again he withdrew from the scene. And then the Independent Television Commission came to its judgement on the licence: 'it did not appear to the Commission that the applicant would be able to maintain its proposed service throughout the period (10 years) for which the licence would be in force' (ITC press release, 18 December 1992). The licence would not be awarded. Thames Television, leader of the unsuccessful bid, now with the backing of Time Warner, has not yet worked out how best to challenge the ITC's decision. Meanwhile, the ITC has begun to reconsider the future of the channel, and one possibility it is considering is a series of local stations in major provincial cities (Burnett 1993). The localist cause has, for the moment at least, a new lease of life.

What became clear, then, in the short prehistory of the new channel, was how much its shaping reflected the underlying dynamics at work in the media industries. The saga of Channel 5 also shows how local and regional agendas are having to accommodate to the forces of globalization. The terms of reference moved, in a very short time, from the interests of provincial cities and regions, to an awareness that the real shapers and shakers in this industry were the global players, and to a recognition that it is in the interests of cities and regions to come to terms with this reality. The possibilities that are perhaps still on offer can therefore only be about some kind of local–global compromise. As yet, the only financially viable kind of 'local' for Channel 5 would seem to be the capital city, the metropolis of London. Whether the compromise could be achieved by smaller cities and regions remains very much in question.

CITIES AND REGIONS IN THE NEW MEDIA ORDER

The historical development of regionalism during the twentieth century has been understood as 'the reaction against a centralising and standardising process': 'Regionalism proceeds from a reaction against the centralised government of the modern state, against the centralising influence of the capital city, and against the standardising effects of modern civilisation' (Gilbert 1939: 68–9). Broadcasting and other media have been seen, over a long time, as making an important contribution in articulating the distinctive interests and identities of local and regional communities, and there are no signs that things will be any different in what we have called the new media order. In these global times, it seems that there is as much as ever a desire 'to re-territorialise the media . . . to establish a relationship between the media and territory' (Lafrance 1989: 294–5; cf. Pailliart 1989). There remains a great interest in how the media can contribute to sustaining the distinctiveness and the integrity of local or regional economies and cultures against the forces of centralization and standardization.

But, of course, what is being reacted against is no longer the same as it once was. As Neil Smith (1988: 150) argues, 'regional differentiation becomes increasingly organised at the international rather than national level; sub-national regions increasingly give way to regions of the global economy'. In the case of local or regional broadcasting initiatives, it is, then, no longer—or no longer simply—a question of reaction against the centralizing and standardizing influences of the national state and culture, but rather of realignment and repositioning in the context of supranational media markets and cultures. It is in this respect that it is appropriate to see these new developments in terms of neoregionalism or neolocalism.

As we have argued, the most significant context for this process of realignment and repositioning has been the European audiovisual space.

What is significant and notable about this changed context is not just the deconstruction of the national framework, but the diminishing significance of a single centre as a point of reference for regional economies and cultures. Localities and regions are tending to define themselves horizontally, as it were, in terms of the multiple and decentred system of European cities and regions. The media initiatives that we have discussed in the previous sections have all held up the possibility of escaping from their subordinate position as peripheral regions in the national territory and of doing this through incorporation into what they see as the potentially more open and even structure of the European regional system. 'Europe of the regions' is a slogan that has strategic and instrumental value in the effort to break the mould of the old hierarchies of the national broadcasting space.

This new European agenda has put a premium on local or regional enterprise and competitiveness. It reflects, of course, a broader process of change that has bestowed—or imposed—greater economic and entre-preneurial responsibilities on subnational units. 'Complex international interdependence and the structural and functional weakness of the nation state invite subnational units to become increasingly involved in direct international activities', and as this becomes the case 'subnational governments are prompted to plan their international activities, to develop paradiplomatic strategies in seeking the most effective way of inserting themselves into the new international economic system (Soldatos and Michelmann 1992: 132, 130). In the case of the broadcasting and audiovisual industries, this imperative brings together a varied coalition of organizations (including local or regional authorities, media development and promotional agencies, and programme-makers) to coordinate local activities and capacities and then to represent their interests at inter-national forums. This international involvement might include lobbying for resources (for example, from the MEDIA programme or the European Social Fund), promoting the region (through attending trade fairs and film festivals), or attracting inward investment (in the form of location shooting). If these developments seem to offer cities and regions new possibilities to control their own destinies, what should be clear is that the mechanisms of paradiplomacy are institutionally fragile and highly competitive. In this new game, too, there are, of course, winners and losers, and consequently there will inevitably be new urban and regional hierarchies, this time at the European and international level.

What should also be recognized is the position of relative weakness from which most cities and regions are competing in the new media order. There are considerable difficulties in bringing any influence to bear on the global firms and organizations which now dominate the broadcasting and audiovisual sector. Local and regional interests are primarily concerned with the production of film and television material. What is absolutely critical in this sector, however, is the distribution function, which ensures access to markets. And what has been striking about the development of the media industries over the past ten years or so has been the rapid concentration that has occurred in this key distribution sector. The nature of audiovisual production also makes it highly dependent on access to development funding and production finance, and again this puts global companies in a gatekeeping position. When this domination by global interests is taken into account, the necessarily reactive and opportunistic nature of regional media development initiatives becomes apparent. The forces shaping the industry remain beyond the control and influence of local actors. The agenda is shaped by the activities of large media companies, and local or regional production industries can only struggle to adapt to the adverse conditions of this environment.

Ironically it is also likely to be the case that whatever success the new types of local or regional strategy do manage to achieve may actually contribute to undermining the stability of a regional broadcasting industry. In so far as media development strategies have necessarily shifted towards the attraction of inward investment and expenditure, they have contributed to the intensification of inter-regional competition. Competition between cities, regions, and countries for inward investment or inward expenditure works to promote the mobility of the industry as a whole, thereby making the task of sustaining or retaining economic activity all the more difficult. It is likely that the net effect of the proliferation of screen commissions and investment funds will be to create a kind of international infrastructure that will only increase the flexibility of large media organizations. This, in turn, will make the business of attracting and retaining the activities of such firms more costly and competitive. The success of such strategies simply takes the race between cities and regions to higher levels.

What, then, of the relation between broadcasting and territory in the new order? What of the contribution of broadcasting to the articulation of local or regional interests and identities? What we in fact see is a conflict, or even a contradiction, between economic and cultural objectives: the economic imperatives driving media development initiatives do not necessarily work towards the consolidation of cultural community and may even work towards its dissolution. The key point is that local producers are not—or not generally—producing for local consumption. They are, rather, producing for national and international markets, and are compelled to produce what those markets want. The question, then, as Simon Frith (1991: 144) argues, is 'whether [a city or region] can be both effectively "represented" culturally . . . *and* compete efficiently in the provision of cultural goods for the national (or international) market'. The answer would seem to be a negative one; in most cases, achieving a competitive position in the expanding marketplace is not compatible with the projection of local or regional distinctiveness.

And if we look at the city or region as a space of consumption, rather than of production, the prospects are equally problematical. Here we must take account of Geoffrey Nowell-Smith's point that the audiences are now considered economically and subdivided in terms of segments of a consumer market. In terms of potential revenues (through advertising, sponsorship, or subscription), localized or regionalized media markets are clearly significant. When it comes to programming, however, the situation is that local markets do not translate into local programming. The economic reality is that, at a local or regional scale, the critical mass of the audience required to pay for quality programming is not attained, and the only option is then to take supply from outside (usually from the US market). This was clearly the case, for example, with the proposals for Channel 5, which included cheap local programming (mainly news), but which depended on Hollywood films and music videos as the real audience grabbers. It has long been the case that the more 'local' the organization of the broadcasting market, the more it is reliant on the product of global suppliers. From this consumption perspective too, then, there can be no straightforward equation between local television and more local programming content.

The idea of local or regional broadcasting becomes easily associated with the ideal of community, and then community becomes associated with ideals of territorial and cultural integrity. What we have outlined as the conflicting and contradictory processes at work in the new media order call into question any such sense of imaginary unity and coherence. The reality of contemporary geographies, as Doreen Massey (1992: 6) has argued, is that the boundaries of place are dissolving, different geographical scales becoming difficult to separate, and geographical diversity is formed 'not so much out of a home-grown uniqueness, as out of the specificity of positioning within the globalised space of flows'. The danger in this new order is that, whilst economic forces are dismantling borders and frontiers, our cultural reflexes will resist the exposure this entails and cling on to the belief in home-grown uniqueness. The most general problem of our time 'is that of the contrast between the opening up of frontiers and nostalgia for closed and stable communities' (Hassner 1991: 20). Local and regional broadcasting, if it is to live with this new reality, must come to terms with the fact that we are no longer 'in' cities and regions in the way we once were—or thought we were.

REFERENCES

Aksoy, A., and Robins, K. (1992), 'Hollywood for the 21st century: Global competition for critical mass in image markets', *Cambridge Journal of Economics*, 16: 1–22.

Bauman, Z. (1992), 'Soil, blood and identity', *Sociological Review*, 40: 675–701.

Boden, T. (1988), *Featuring the City: American and German Experiences of the Location Business* (Birmingham: Birmingham Film and Television Festival).

—— (1990) (ed.), *Cities and the Media: The Location Business* (Birmingham: Birmingham Film and Television Festival).

—— (1991) (ed.), *Film Commissions and the British Film Industry* (Birmingham: Birmingham Film and Television Festival).

Bourdon, R. (1992), 'Le Programme de télévision et l'identité nationale', *Médiaspouvoirs*, 28: 5–13.

Briggs, A. (1975), 'Local and regional in northern sound broadcasting', *Northern History*, 10: 165–87.

Brooks, R. (1993), 'UK to let as Hollywood is lured on to location', *Observer* (31 Jan.).

Burnett, C. (1993), 'Regional option for C5', *Broadcast* (5 Feb.), 10.

CEC (1986), 'Television and the audio-visual sector: Towards a European policy', *European File*, 14/86.

Channel Five Holdings (1992), *Application to the Independent Television Commission for the Channel 5 Licence: Summary for Public Information* (London: Channel Five Holdings Limited).

Cluzel, J. (1992), 'L'Audiovisuel à la veille du marché unique', *Revue Politique et Parlementaire*, 959: 44–51.

Collins, R. (1992), 'Unity in diversity? The European single market in broadcasting and the audiovisual 1982–1992', Paper presented to the PICT National Conference, Newport, 23–5 Mar.

Cornford, J., and Robins, K. (1992), 'Development strategies in the audiovisual industries: The case of North East England', *Regional Studies*, 26: 421–35.

Fisher, M. (1987), 'Whose cultural industries? A strategic approach', in T. Boden (ed.), *Cities and City Cultures: Local Authorities and the Cultural Industries* (Birmingham: Birmingham Film and Television Festival).

Frith, S. (1991), 'Knowing one's place: The culture of the cultural industries', *Cultural Studies from Birmingham*, 1: 134–55.

Gilbert, F. W. (1939), 'Three aspects of regional consciousness', *Sociological Review*, 31: 68–88.

GLC (1985), *The London Industrial Strategy* (London: GLC).

Guback, T. H. (1974), 'Cultural identity and film in the European Economic Community', *Cinema Journal*, 14: 2–17.

Hall, S. (1992), 'The question of cultural identity', in S. Hall, D. Held, and T. McGrew (eds.), *Modernity and Its Futures* (Cambridge: Polity).

Hassner, P. (1991), 'L'Europe et le spectre des nationalismes', *Esprit* (Oct.), 5–22.

Hjarvard, S. (1991), 'Pan-European television news: Towards a European political public sphere?', Paper presented to the Fourth International Television Studies Conference, London, 24–6 July.

Hurd, G. (1987), 'The cultural industries and local economic development', in T. Boden (ed.), *Cities and City Cultures: Local Authorities and the Cultural Industries* (Birmingham: Birmingham Film and Television Festival).

IBA (1989), *The IBA's Response* (London: IBA).

Lafrance, J.-P. (1989), 'Le Territoire médiatique du local', *Netcom*, 3: 293–309.

Levitt, T. (1983), *The Marketing Imagination* (New York: Free Press).

Maggiore, M. (1990), *Audiovisual Production in the Single Market* (Luxemburg: Office for Official Publications of the European Communities).

Massey, D. (1992), 'A place called home?', *New Formations*, 17: 3–15.

Masters, N. (1992), 'Countdown to chaos', *Broadcast* (10 July), 18–19.

Meyrowitz, J. (1989), 'The generalized elsewhere', *Critical Studies in Mass Communication*, 6: 326–34.

Morley, D., and Robins, K. (1989), 'Spaces of identity: Communication technologies and the reconfiguration of Europe', *Screen*, 30: 10–34.

Murray, R. (1991), *Local Space: Europe and the New Regionalism* (Manchester and Stevenage: CLES and SEEDS).

Musso, P. (1990), *European Regions and Television* (Nord Pas-de-Calais: Council of Europe/Conseil Régional).

—— and Pineau, G. (1989), 'La Télévision entre l'état et le marché', *Médiaspouvoirs*, 14: 119–30.

Nowell-Smith, G. (1991), 'Broadcasting: National cultures/international business', *New Formations*, 13: 39–44.

Obuchowski, J. (1990), *Comprehensive Study of the Globalization of Mass Media Firms* (Washington, DC: National Telecommunications and Information Administration, US Dept. of Commerce).

Pailliart, I. (1989), 'De la production des territoires', *Médiaspouvoirs*, 16: 58–64.

Pegg, M. (1983), *Broadcasting and Society, 1918–1939* (London: Croom Helm).

Pocock, J. G. A. (1991), 'Deconstructing Europe', *London Review of Books* (19 Dec.), 6–10.

Robins, K. (1989), 'Reimagined Communities?', *Cultural Studies*, 3: 145–65.

—— (1993), *Geografia dei media: Globalismo, localizzazione e identità culturale* (Bologna: Baskerville).

—— and Cornford, J. (1991), 'Looking local', *Marxism Today* (Nov.), 38–9.

—— —— (1992), 'What is "flexible" about independent producers?', *Screen*, 33: 190–200.

Ross, S. (1990), 'Worldview Address', delivered at the Edinburgh International Television Festival, 26 Aug.

Rowley, C. (1990), 'Channel 5: Can it make viewers happy—and make money?', *Television: Journal of the Royal Television Society* (Jan.–Feb.), 9–15.

Scannell, P., and Cardiff, D. (1991), *A Social History of British Broadcasting*, i. *1922–1939, Serving the Nation* (Oxford: Basil Blackwell).

Schiller, H. (1991), 'Not yet the post-imperialist era', *Critical Studies in Mass Communication*, 8: 13–28.

Sepstrup, P. (1989), 'Implications of current developments in Western European broadcasting', *Media, Culture and Society*, 11: 29–54.

Smith, N. (1988), 'The region is dead! Long live the region!', *Political Geography Quarterly*, 7: 141–52.

Soldatos, P., and Michelmann, H. J. (1992), 'Subnational units' paradiplomacy in the context of European integration', *Journal of European Integration*, 15: 129–34.

Trelluyer, M. (1990), 'La Télévision régionale en Europe', *Dossiers de l'Audiovisuel*, 33: 10–55.

11

Global–Local Social Conflicts: Examples from Southern Europe

Costis Hadjimichalis

INTRODUCTION

A few years ago it may have sounded anachronistic, not so radical (or Marxist) or even wrong, to insert notions such as place and local into the analysis of the uneven geography of capitalist development. On the one hand there was the tradition of community studies and that of places for 'their own sake' which justified mistrust and criticism. On the other, the preoccupation of most radical and Marxist studies with structural or systemic processes of accumulation, and a focus on 'wider systems', reduced places and localities to simple arenas of production and reproduction.

Following from more recent radical research on economic and socio-spatial restructuring around the globe (in which the feminist critique has played a decisive role) places are back in the agenda of social science. Not as idealized communitarian entities and simple locations in a worldwide accumulation chain, but as loci of particular social relations, of unique and at the same time general trajectories, where men and women struggle through everyday life producing and consuming products and culture for and from world markets.

To be sure there is major disagreement on whether this change of research agenda is a regressive return to empiricism or a move forward. There is also disagreement on whether local areas—with the exception of a few paradigmatic cases which the literature tediously repeats—can cope with or survive in intensified global competition.

The debate still goes on. Despite their disagreements, commentators have come to take local areas and places more seriously than before, to consider them in a global framework as parts of globalized processes, and arguing, following Massey (1991), for a 'global sense of place'. However, what has been missing from the discussion so far is an explicit acknow-ledgement that definitions of local, place, and global are not fixed, but contingent upon conflicts over the control of political power, resources, space itself, and cultural identity. The definition of global or local for any society or social group is not given, but depends on conflicting interests based on class, gender, ethnicity, religion, employment, or geographical origin. The meaning of global and local is what historical actors struggling in such society and place try to make of it. In many cases global means simply 'the outside world', which itself is part of what constitutes a place. Depending on the degree of inclusion–exclusion of each place in the world economy, 'global' can refer to the next city and region or the actual global economy and society.

The interaction between global and local processes includes exploitative relations. Territorially specific social groups struggle with each other and against their own state or supranational political institutions such as the EC, or against de-territorialized economic interests such as multinational corporations. In this chapter an attempt is made to analyse the interests and actions of territorially based socio-political movements particular to a place, and which have been generated from a local–global conflict against changes to place-bound social relations. The argument will be illustrated with cases from southern Europe, where territorial mobilizations were important elements in the construction of local civil society until the beginning of the 1980s. Since then, they have declined following a general trend of de-politicization.

THE NEGLECT OF SOCIAL ACTORS AND SOCIAL ACTION

Radical literature on uneven regional development and the relation between local and global processes focuses primarily on issues such as capital restructuring, global flows, transnational corporations, financial structures, and the like. A few paradigmatic cases of local capitalist success in high tech parks, old and new industrial districts, rejuvenated central cities, and so on, have allowed the construction of a positive image of the local–global link (Sabel 1989). Similarly, in the neo-corporatist literature, the emphasis is on local and regional business associations and how they manage to sell their place-bound uniqueness in global markets in terms of products, quality, labour skills, historical heritage, and so on (Coleman and Jacek 1989).

In both approaches, local capitalist success in a global competitive framework has been analysed from the point of view of the firm and its technological, institutional, and social environment. It seems that from an early neglect by radical geography of some positive aspects of market competition and inter- and intra-firm organization, we have shifted towards fetishizing everything that has a firm-oriented analytical focus or emphasizes post-Fordist, post-modern, and flexible tendencies in the economy (Sayer 1992; Chronaki *et al.* 1993). I am not denying the importance of corporate, market, and institutional analysis, but what strikes me is the underestimation at present of the literature on the local–global nexus in terms of the role of people and their organizations as social agents affecting change.

In the 1970s there were some interesting studies in this tradition within

French and Italian sociological and political literature (see Dulong 1978; Bleitrach and Chenu 1979; Castells 1977; Cottone 1972; de Vos 1975), but since then relatively little attention has been given to local struggles and social movements defending or demanding improvements for their own place (some exceptions, however, are Thrift 1983; Doherty 1983; Rokkan and Urwin 1983; Bourdieu 1980; Pickvance 1986 and forthcoming; Hudson and Sadler 1986). This is a notable omission, given well-developed arguments in the literature on urban social movements. The omission might stem of course from an antilocalist tradition within the Left, which correctly has identified and criticized cases of reactionary localism, place-boosterism, or petty-bourgeois local protectionism and xenophobia. To this we have to add that early Marxist research on nationalism and regionalism was trapped in between a utopian class internationalism and an economism, which prevented it from giving a historical account of territorial struggles and social movements other than as class struggles at the point of production (Haupt and Weill 1974). What is needed therefore, as Massey (1991) argues, is the development of 'a progressive sense of place' in the context of current global restructuring, one in which place can be seen, as Alan Pred (1989) wishes, as 'a potential side of struggle and resistance'.

To defend a place from global penetration or to promote some of its qualities is not necessarily reactionary or static. This depends on the type of strategy and issue. It makes a difference, for instance, whether the coastguard patrols the Mediterranean to 'defend' EC borders from poor Third World migrants (sometimes by throwing them into the sea) or whether people defend an ecologically important lagoon from the pollution of multinational chemical investments. It makes a difference whether peripheral areas compete with each other to 'promote', that is sell immodestly, local qualities (monuments, food, beaches) to global tourists for hard currency, or whether they struggle to stop 'adaptation' of the local building laws to meet the demands of international hotel chains. And, to see local–global interaction from another angle, there are some known cases in which local entrepreneurs in 'successful' industrial districts have come together to refuse to adapt to national and EC regulations on illegal labour and social security contributions for homeworkers. In all these examples there has been a conflict between some aspect of global process and a local priority. The question is how to preserve place uniqueness *vis-à-vis* global intervention (or at least what appears as global) without ending up in a parochial or reactionary position.

Different Forms of Territorial Mobilization

There are a number of ways in which we can approach territorial mobilization. I may begin by pointing out first, that territorial mobilization can be built upon many, often unpredictable, parameters among which we can identify three which appear most frequently (see also Rokkan and Urwin 1983): cultural and community identity; material, social, environmental, and economic initiatives; and political reasoning and expectation. These three parameters are mediated through place, class, gender, and ethnicity, and together may provide a basis for mobilization. Second, a local or regional movement is justified not so much because it is based in particular localities, but because it introduces place as a basic component of socio-political organization and action. In other words, mobilization may be legitimized at the territorial level when it is a social response to a crisis of territorially specific goals related to the parameters mentioned previously, triggered by a particular model of economic development and political regulation. In this sense a territorial mobilization may rise out of a particular place. It may refer to local symbols and meanings; to conditions of living, learning, and working; to the physical and natural environment in a given place; and it may oppose global crisis and restructuring affecting that given place.

Although working-class areas and areas of localized poverty and socio-economic disintegration have traditionally had greater impetus for political struggle, territorial mobilizations have arisen in a diversity of areas and involving a variety of social groups. Therefore, they cannot be considered as a mere extension of the class struggle between capital and labour, or only representative of the deprived, the marginal, and the underprivileged. Alain Touraine in his book *Voice and Eye* defines social movements as 'socially conflicting behaviours but also culturally oriented forms of behaviour and not only as the manifestation of objective contradictions of the system of domination' (1981: 78). A territorial movement, thus, need not change everything for itself. However, it should attempt at least three things: greater access to bases of social power; a change in social identity; and a reduction of conflicts between local and global processes as well as those in local relations of production and collective life.

Here, particularly helpful is the distinction between class and status-group interests and conflicts (see also Arrighi, Hopkins, and Wallerstein 1989). Class conflicts (intra-capitalist and/or capital–labour conflicts) emerge as a result of antagonistic relations in economic and political arenas, either directly in relation to other classes, or indirectly, through their relation to the political apparatus (e.g. the state or the EC). Status-group conflicts emerge as a result of collective interests which are impossible to accommodate into class conflicts. Both types of conflict are justified theoretically and empirically, to account for collective action or for its absence. Under certain conditions either class or status-group conflicts, or both, may acquire a territorial dimension. This may happen when a place-based consciousness is built along class and status-group interests identifying with a specific territory.

Territorial conflicts might also emerge from global–local interaction also as the product of historical changes in the spatial division of labour. Conflicts are associated with the legacy in places, of divisions of labour of earlier periods of capitalist development. Localities and regions defined in this way are thus the product of an ongoing conflict which is, as Soja (1989: 168) argues, 'tentative, ambivalent and creatively destructive as any other component of the spatial matrix of capitalist development'. This 'creatively destructive' conflict may give rise to territorial mobilization around certain issues.

A Typology for Southern Europe

In southern Europe, for example, territorial mobilizations shortly before and after the fall of various dictatorships became important components of everyday politics and played a major role in shaping local civil society and the development policies of new governments. There are, however, of course, many forms of territorial mobilization as well as conflicting explanations of their origins and importance, some of which are discussed below following a simple typology. Although all definitions of the term are not covered, the aim of the typology is to facilitate comparative analysis, and to illustrate how my initial hypothesis of global–local conflicts fits with southern European reality in the last two decades (see Table 11.1).

One important category comprises those historical movements combining ethnicity, localism, and strong local socio-cultural identification with economic and political goals. Well-known cases in southern Europe include the Basque movement, the autonomy movement in Catalonia, and the Corsican movement. Less-established and less-organized movements exist also in Galicia, Andalusia, the Canary Islands, and the Azores, while two minor pockets in north Italy are the Friulians and Ladins, both speaking dialects of Rhaeto-Romanic (de Vos 1975). Their long-term aims and practices focus on federalism or regional autonomy, and they vary in tactics, ranging from militant struggles to bids for parliamentary institutional changes.

Another category comprises those cases which have been generated from a local–global conflict, and are based on *ad hoc*-issue coalitions serving as a basis for territorial mobilization. One manifestation is local or regional social movements, composed of strong territorial alliances between classes and status-groups, defending or demanding improvements to local conditions. In this case the movement relies on strong local institutions and has a longer duration than the actual conflict. Another manifestation is local or regional protests, where people may struggle on a temporary basis, defending a place or putting the demands of a peripheral area on the agenda of the central state political system. In both cases, actions may vary along a spectrum ranging from direct confrontation to peaceful written petitions.

TABLE 11.1. *A typology of territorial movements*

	Social protests		Social movements	
	Integrated	Unintegrated	Integrated	Unintegrated
Offensive	Crete (1979)	Grevena (1980)	Pylos A (1974)	Alentejo (1975–7) Emilia-Romagna (1962–78)
Defensive	Friuli-Venezia (1968)	Cantabria (1985–6) Kalamas river (1990)	Brescia (1977) Sicily-Calabria (1987–90)	Pylos B (1977) Andalusia (1978–84)

Finally a distinction can be made in terms of integrated and unintegrated mobilizations (see also Lankowski 1987), and we may further distinguish within each of these forms, their aims from their tactics. In the former, when aims pursue conflict solutions within the system, i.e. without breaking with existing rules, we may call them an integrated protest or movement. The opposite is true when aims are against established rules and decisions. In the latter case, as shown in Table 11.1, we can distinguish between defensive (i.e. against a particular intervention which could threaten certain local gains) and offensive tactics and actions (i.e. demanding certain changes and improvements of local conditions of various sorts).

TERRITORIAL STRUGGLES IN SOUTHERN EUROPE

For southern European people, the mid-1970s were a dramatic turning-point in their recent history. For Greeks, Spaniards, and Portuguese the fall of dictatorships was strongly related to fundamental economic and social transformations which, in many instances, brought about changes as important as those that took place at the political level. For Italians the industrial prosperity of previous decades had slowed down and exposed its internal limits. Such wider conditions served to reveal the accumulated layers of problems within different localities, and to give expression to them as territorial conflicts, thus leading to different local or regional mobilizations based on the nature of the articulation between local demands and global processes. One could argue that Italian territorial movements should be treated separately from those in Greece, Portugal, and Spain after the end of dictatorial rule. The logic behind a common comparative schema derives from the existence of a strong left-wing or communist social base in all these countries as well as common bitter internal struggles within each movement for leadership. I must also clarify from the outset that the social protests and movements that I will discuss below are by no means unique, nor are they necessarily the most important within southern Europe. Other places across Europe have experienced similar mobilizations, while due to lack of adequate documentation many interesting cases are not included in this discussion.

A local or regional integrated protest is generated when a certain intervention produces a conflict between two territorial hegemonic blocs (or with some non-local political or economic entity) with minor popular participation. The term territorial hegemonic bloc is a concept derived from the work of Antonio Gramsci (1975) and Nicos Poulantzas (1978) and consists of: a specific local system of exploitation and articulation of various forms of production; a specific form of political and ideological mediation towards dominated classes and status-groups; and a specific form of alliance among dominant classes and status-groups under the hegemony of a local bourgeoisie, if one exists (Lipietz 1985; Hadjimichalis 1987).

Two examples of such integrated protest were in Friuli-Venezia Giulia in 1968 and Crete in 1979. The first was a defensive protest generated out of the conflict between certain fractions of the regional bourgeoisie and the Italian state and the EC, around planning proposals (Arcangeli 1982). At that time the regional hegemonic bloc was against the introduction of extended hand-use regulation which could threaten established networks of small- and medium-sized enterprises (SMEs) in Christian Democrat-run industrial districts. The second was an offensive one, in which business associations in Crete pushed the Greek state in the late 1970s to concede additional subsidies to combat competition from other countries in the EC Mediterranean, especially in agriculture and tourism. Among the major demands was the acceptance by local and EC authorities of the use of low-paid and non-declared foreign labour employed seasonally in the fields and in various tourist ventures.

These conflicts between either territorial hegemonic blocs or one of the latter and the national state or the EC, do not attract popular participation, but enjoy strong support from local institutions. Business associations, local chambers of commerce, and the like have always supported, if not initiated, such mobilizations, trying to present their differences as those of the popular masses.

In cases of unintegrated local or regional protest, in contrast, people have mobilized without the support of local institutions, and only in sporadic collaboration with local SMEs. Two cases of defensive mobilizations are the protest of small milk-producers in Cantabria in the mid-1980s and more recently, protests against intrusive uses of the Kalamas river in Epirous (Greece). In Cantabria (Spain) the imposition of EC limitations on milk-production generated a strong regional protest in 1985–6 leading to a three-week regional strike which succeeded in stopping the construction of a big Nestlé factory in the area. Small milk-producers were also mobilized against the National Basque party of the local chamber of commerce which did not support their demands. In Kalamas river, local people and small businesses in 1990 went on strike, blocked main roads, and occupied public buildings in Igoumenitsa (Epirous) for two months, protesting against a state decision to use the Kalamas river as a waste disposal facility for the city of Ioannina. The mobilization attracted huge publicity from the media and has succeeded until the time of writing in stopping the project.

In Grevena in northern Greece an offensive regional protest was initiated against regional economic backwardness. Organized during the winter and spring of 1980, it started with a four-day regional strike in which regional economic life practically stopped. The major slogan of the protest was 'Jobs and social security for our region. Do not force us to migrate.' A social bloc composed of small entrepreneurs, workers, and peasants, it elected a coordinating committee, which presented a four-point declaration to the government and to the EC Regional Development Fund, demanding improvements in regional means of collective consumption, job security, and new investments (Hadjimichalis 1987).

If a territorial hegemonic bloc succeeds in securing not only passive support by dominated classes and status-groups, but also their active mobilization beyond the period of actual conflict, then this can be referred to as an integrated local or regional social movement. An interesting example of offensive mobilization is the case of Pylos in Greece, where in 1974 a multinational company generated support from an entire local social coalition for its investment proposals for a huge port-industrial complex. The mobilization was opposed by the Ministry of Culture and the Ministry of Planning and the Environment, which denied permission for the support because of archaeological and environmental reasons. Our research in the area showed that the multinational company was supported mainly by petty bourgeois landlords and right-wing deputies, who were able to mobilize local people with the promise of future industrial employment. Three years later, when local people became better informed about the limited employment possibilities and the severity of potential environmental destruction, an oppositional local movement grew, but this time it was unintegrated. In the first movement local institutions related to commerce, agricultural cooperatives, and tourism supported the multinational company on the basis of expected place-boosterism. In the second, unintegrated and defensive mobilization, none of the established local institutions were initially supportive, but after a year some local authorities and two small environmental associations joined the campaign. However, owing to wide publicity in the national media, the defensive movement found support from different parts of Greece (Hadjimichalis and Vaiou 1980; 1990).

Another interesting defensive regional social movement was that generated in 1977 around steel closures in the region of Brescia in Italy. To cope with the global crisis of overproduction in the steel industry from 1975 onwards, the Commission increasingly intervened in the market through setting prices and quotas. In 1977 it stopped the advantage of the 'Bresciani'—small steel-producers in the Brescia region—who then were able to offer concrete reinforcing bars much more cheaply than their EC competitors because of the deployment of a modern electro-steel process and lower wages. For one year a massive integrated regional social movement was mobilized, involving both the regional business association of steel-producers and the workers' union. The movement at the end, as often the case, succeeded in gaining high compensations for the industrialists but limited ones for the workers.

Under certain conditions the interests of dominated classes and status-groups can coincide temporarily with those of a fraction of a territorial hegemonic bloc. This has been the case in Sicily and the southern Italian mainland since 1987, where local people have made an alliance with a fraction of the regional bourgeoisie and the state to fight against the domination of the Mafia. Despite its limited success this coalition is a major social development in itself, if one takes into account the pervasive regional and local rootedness of the Mafia in the Mezzogiorno. Several protest rallies have taken place, one of the largest of which was in September 1989 when more than 300,000 people marched in Palermo.

Finally, unintegrated territorial mobilizations might take place when proposed changes in the spatial division of labour generate a conflict between local dominated classes or status-groups and a global force. In this case territorial reaction is focused either against local élites who have allied themselves with exogenous forces, or directly against the latter. It is this form of territorial mobilization, namely the rise of popular local or regional consciousness around local issues, which I consider the prime example of an unintegrated territorial social movement. It constitutes a type of social practice at the local or regional level drawing upon social classes and status-groups who identify with a particular place, contradicting the institutionally dominant social logic and challenging global decisions which affect local uneven development.

An example of an unintegrated defensive movement is the case of Andalusia in Spain during the late 1970s. Initially a petty-bourgeois–small-peasant coalition, it then made an alliance with working people to form a joint committee against restructuring processes in agriculture threatening to result in massive unemployment and greater concentration of landownership in the Guadalquivir valley (Titos Moreno and Rodriguez 1979). This committee evolved into an important local institution influenced by the regional Socialist party, but also including anarchist militants. In 1984, however, with the Socialist party in office, Madrid blocked the long-awaited land reform toward which some progress had been made over the preceding years. This was a direct response to pressure from the European Commission which wished to cut back the already large agricultural surpluses of the Community. A three-month regional mobilization, with violent confrontations with the police as well as protests in Brussels, failed to change this decision.

Good examples of offensive forms of unintegrated regional social movements are found in the rural Alentejo (southern Portugal) and the Emilia-Romagna Development Coalition. In Alentejo one of the most radical mobilizations in southern Europe occurred from 1975 to mid-1977, under the leadership of the regionally strong PCP (Portuguese Communist party). Within a few months of the Portuguese revolution, the *latifundiarios* (absentee landlords) who dominated the region were virtually abolished, and the power structure of the wheatlands of southern Portugal was completely altered. By the end of 1975 480 estates covering 2.5 m acres (one-fifth of Portugal's farmland) had been occupied and transformed into cooperative farming, creating 10,000 new jobs (Gallagher 1986).

In the region of Emilia-Romagna in Italy, after 1962, with the Communist administration often at the helm, a development coalition was established based on four goals: radical democracy, decentralization, local development based on SMEs, and popular consensus (Zangheri 1975). The coalition involved many local institutions, which supported a four-year development plan (*Piano Programma*), and new legislation (passed in 1974) for urban and industrial development, under which 500 neighbourhood councils were established.

FROM PROTEST AND MOBILIZATION TO MANAGEMENT

In southern Europe territorial mobilizations arose during the 1960s and 1970s generally within a radical political framework. As they increased their importance, so too did their need for legitimation at an administrative level and, more specifically, for positive government action. The high degree of economic and ideological upheaval within the four nation states concerned made legitimation of the movements indispensable owing to the radicalization of the local populations. The latter were no longer prepared to restrict themselves to purely economic demands, since these could not be met within the framework of existing national economic inefficiency. The result was an authentic crisis of hegemony. The issues of uneven regional development, political conflict, cultural distinctiveness, and environmental protection became the focus of social protest.

The situation, however, changed considerably during the 1980s and early 1990s. It was not only the global economic crisis which affected the societies of southern Europe. More important were changes in their internal social and spatial structure. From the perspective of uneven development, regional inequalities still persist, but a process of rural transformation and productive decentralization away from old industrial centres has changed long-established dichotomies between cores and peripheries, between 'north' and 'south' (see Chapters 3, 5, and 6; Bagnasco 1984; Paci 1982). Capitalist development started to flourish spontaneously in regions and localities whose economic performance, based upon a mixture of intensive agriculture, small industry and tourism, social division of labour (formal and informal activities), and degree of state intervention, has become characterized as an 'intermediate' level of development (Garofoli 1992; Vasquez-Barquero 1992; Ferrao 1987; Hadjimichalis and Vaiou 1990). Many of these intermediate regions had hosted active regional and local mobilizations only a few years previously (e.g. Crete, Galicia, Emilia-Romagna, Andalusia, and the Veneto).

At the political level, the most important change was the coincidence during the 1980s of simultaneous victories of socialist parties across southern Europe. Soon, however, it became clear that while their apparently radical left programmes were sufficient to mobilize support for electoral victory, they were inadequate and superficial as a basis for effective domestic policies. In the end, and despite their rhetoric, the proposed 'socialist' policies (including local and regional ones) differed little from an updated Keynesianism with neopopulist slogans (as in Greece), or even a neoliberalism with pseudo-socialist rhetoric as in Italy, Spain, and Portugal (Petras 1984).

The crucial problem with these socialist parties (in many cases also communist parties) has been their relation with mass movements and especially territorial mobilizations. The Left in southern Europe (with a few exceptions) has been unable to understand these new forms of social mobilization and has been unprepared to accept their autonomy. The basic argument of the national, regional, and local leadership has been, and still is, that its strategy of 'change' automatically incorporates all social movements. This party-integrationist ideology draws upon a conceptualization of regional and local issues as simply one step towards a national electoral strategy. Thus, similar to the history of social movements elsewhere, in southern Europe, whenever a party (usually but not exclusively of the Left) tried to 'guide' them, the movements disintegrated or dissolved.

Economic changes resulting from productive decentralization at the local level, and political changes due to the rise of socialist parties after a long period of domination by right-wing authoritarian regimes, permitted new forms of social mobility which found expression in another political development, namely the rise of middle-class predominance in local and regional politics, and the rise of a managerialist local ideology based upon efficient management. These changes are even more pronounced in the cases where attention focuses narrowly upon movement leaders and activists. Thus, in many areas, relative economic improvement and political change permitted middle-class activists to guide territorial struggles towards populist, non-class, positions, yielding disintegrating effects in the long term.

Under these conditions it is quite clear why the situation by the end of the 1980s was different from the mass mobilizations of the 1960s and 1970s. Today some of the demands of the earlier integrated movements and protests have been met, while other 'unintegrated' demands have been either postponed or resisted by the nation state, by supranational organizations such as the EC, or by multinational capital. In Greece, in a few cases, local environmental protests and movements succeeded in opposing coastal destruction from highly polluting global industries. Other local popular uprisings, however, defending old industrial localities from global restructuring have been defeated (e.g. Lavrion, North Evia, cooperative food industries in Thraki). Crete and Grevena now have their own respectable development plans which have replaced activism and spontaneous bodies, while a continuous rhetoric centred on decentralization and popular participation introduced by the socialist party (PASOK) after 1981 has led many to believe that the state and the EC now take regional problems seriously.

In Portugal, the radical experiment in Alentejo has been undermined following a law passed in 1977 which permitted large parts of the land to be returned to the old *latifundarios*. This was the result of a coalition of interests between the powerful landed class, EC regulations, and clientelist local authorities controlled by the Portuguese communist party (Gallagher 1986). It is not accidental that one of the arguments put forward by Christian Democrats and Socialists in Portugal in support of EC membership was that accession would necessitate the de-collectivization of farms in Alentejo and a more liberal policy towards foreign investment. Today the region has a Regional Development Corporation managing it from Lisbon.

In Emilia-Romagna the development coalition today acts less as a radical, self-managed alternative than as an autonomous bureaucratic institution of the PDS (Partito Democratico della Sinistra, which succeeded the Communist Party in 1990), which sells to world markets the region's comparative advantage drawing upon specialized industrial districts, agriculture, and public monuments (Nardin 1987). It has also opened, like other regions, an office in Brussels, while the regional federation of SMEs (controlled also by the PDS) demands EC regulation to 'discipline' local labour and to legitimize 'lavoro nero' from Third World migrants (i.e. unregistered labour).

Recent federalist regional movements (the *leghe*) in Italy echo its old nationalistic cultural tendencies. Mobilizations in the north, especially in Trentino Alto Adige, Piedmont, and the Veneto (1986–8), now demand more regional autonomy and a stop to or elimination of migrants from the Mezzogiorno. Their political ideology is ambiguously on the extreme Right, as revealed by their slogan: 'We are neither on the Left, nor on the Right or Centre; we are in the North.'

These explicitly racist regional mobilizations have been highly successful in recent regional and national elections. In the course of the 1990s, Lega Lombarda, a right-wing regional party in Lombardy, has succeeded in winning first place, leaving the Socialists and Christian Democrats well behind. Its campaign has been based on a vigorous critique of all the national parties for their inefficiency and corruption. The Lega has been explicitly regionalist and pro-EC (with slogans such as 'Lombardy for Lombardians') and implicitly racist, favouring the exclusion of all foreign workers, including southern Italians. Lega Lombarda, as well as other *leghe* movements in Italy, have their social base among young urban technocrats, small entrepreneurs, small farmers, and self-employed artisans. These are social groups which were successful during the 1980s and see themselves today as victims of state inefficiency, prevailing political corruption, and unproductive public spending on the 'backward' south and the old industrial areas of the north. Global restructuring and prospects of further European integration have given rise to separatist expectations on the part of prosperous northern and central regions, via membership of the 'European Core' (i.e. the zone of the German Deutschmark), leaving behind the rest of 'backward' and 'corrupt' Italy, whose labour is no longer required.

CONCLUSION

Amidst a certain pro-market and pro-firm euphoria in academia, it may sound banal and 'modernist' to argue again for the need to study socio-spatial inequalities, and social conflict and social action in space. Integration tendencies at two extreme poles, the global and the local, will not eliminate these inequalities, but may give rise to more intense territorial conflicts and movements. The need to problematize the local–global relationship, particularly its conflicting aspects, is essential in order to understand the new dynamic of world capitalism. Alain Touraine wrote in 1978 that 'social movements are not dramatic and exceptional events. They are, in a permanent form, at the very core of social life' (1981: 28). If this is so, three open questions requiring further discussion are relevant.

The first is why some people in particular places are mobilized to challenge an imposed global–local link, while others remain silent. From the few cases presented here we cannot 'prove' any general law or tendency, and I wonder whether this can ever be achieved, since territorial social movements are not mechanical or predictable reactions to historical stimuli—they come and go as people through their actions try to fundamentally modify the role of a locality or region in global society. Yet this social aspiration is confronted by other aspirations, for example, efforts to make the local a bureaucratic locus or a site for global investment, a place for consumerism or hard work, a place of better living conditions, or a place from which you simply must escape.

This brings me to the second question, which relates to the progressive nature of each territorial movement. While local social action and mobilization do matter, their direction is not unambiguous, and as many of the examples earlier demonstrated, they are not always to the benefit of those in need. Excluding the obviously reactionary cases such as the *leghe* movements in Italy, more ambiguous cases such as bids to defend jobs or an important environmental asset in a particular place are not necessarily progressive. A local victory may occur at the expense of other places, since all places are linked together in a global competitive framework which does not promise rewards for all. This poses a serious dilemma for any territorial movement which yearns for more of the old leftist class internationalism, instead of the disintegrated plurality of today. Equally important is the internal organization and the aims of each movement. Without under-estimating the importance of struggles over jobs, working conditions, the means of collective consumption, the environment, and the like, a movement which succeeds in securing them may also contribute indirectly to the reproduction of old modes of exploitation and domination between classes, genders, ethnic groups, and religions. For a movement to be considered progressive, it must challenge and consciously attempt to overcome these older and deeper divisions in its territorial actions. The deconstructionist critique of feminists and ecologists is particularly helpful here, not simply because it adds new parameters of social struggle but because it asks new questions and provides new models for collective action.

The last question concerns the causes of the decline of territorial protests and movements. Leaving aside cases of success and those manipulated by capital and/or the state, I wish briefly to address the role of left-wing political parties. In some cases the relationship between a movement and a party has been a productive one. Local mobilizations have strengthened the social basis of left-wing parties as they challenged decisions of the state and/or global capital. In other cases, political parties provided the organizational experience and militancy needed to address major territorial issues, and offered the bridge to oppose the policies of an authoritarian state and global economic interests. In the majority of cases, however, 'local politics', for the parties of the Left in southern Europe (as elsewhere), was considered a mere step towards 'real' power, defined in terms of access to government and the sweeping together of territorial social movements. The underestimation of the dynamics of local civil society, and indeed the tendency to ignore issues of local autonomy had important negative consequences for local mobilizations, left-wing parties themselves, and the socialist project in general. These domestic failures, together with the global crisis of left-wing politics, have discredited radical socialist alternatives. This is one of the reasons why traditional left-wing forces in southern Europe are at present unable to intervene and provide alternatives at the local level. This leaves an empty space which may be filled by reactionary forces, the most typical examples of which are again after fifty years fascist and racist in nature.

Often what remains from local struggles is a series of scattered fragments—some programmes, many different grass roots groups, un-attached symbols. To amount to something more complete, local protests and movements must move beyond the fragments and seek to forge links between local demands and a new, socially more encompassing internationalism, as well as provide a serious challenge to old and new modes of exploitation.

NOTE

In an earlier version this paper benefited from comments by Phil Cooke and Chris Pickvance, and in its present form, from comments by Ash Amin, Ray Hudson, Doreen Massey, and Dina Vaiou. My thanks to these friends, and as usual, I bear full responsibility for the final outcome.

REFERENCES

Arcangeli, F. (1982), 'Regional and subregional planning in Italy', in R. Hudson and J. Lewis (eds.), *Regional Planning in Europe* (London: Pion).

Arrighi, G., Hopkins, T. K., and Wallerstein, I. (1989), *Antisystemic Movements* (London: Verso).

Bagnasco, A. (1984), *La costruzione sociale del mercato* (Bologna: Il Mulino).

Bleitrach, D., and Chenu, A. (1979), *L'Usine et la ville: Luttes régionales, Marseille et Fos* (Paris: Maspéro).

Bourdieu, P. (1980), 'Identité et représentation: Éléments de critique pour la notion du région', *Actes de la Recherche en Sciences Sociales*, 35: 63–72.

Castells, M. (1977), *The Urban Question* (London: Edward Arnold).

Chronaki, Z., Hadjimichalis, C., Lambrianidis, L., and Vaiou, D. (1993), 'Diffused industrialisation in Thessaloniki: From expansion to crisis', *International Journal of Urban and Regional Research*, 17/2: 178–99.

Coleman, W., and Jacek, H. (1989), 'Capitalist collective action and regionalism: An introduction', in eid. (eds.), *Regionalism, Business Interests and Public Policy* (London: Sage).

Cottone, E. (1972), *Riorganizzazione capitalista e lotta di classe nelle campagne* (Rome: Editori Riuniti).

de Vos, C. (1975) (ed.), *État et mouvements régionaux*, Proceedings, IXme Colloque de l'Association Internationale des Sociologues de Langue Française (Paris and Geneva: Mouton).

Doherty, J. (1983), 'Racial conflict, industrial change and social control in post-war Britain', in J. Anderson, S. Duncan, and R. Hudson (eds.), *Redundant Spaces in Cities and Regions?* (London: Academic Press).

Dulong, R. (1978), *Les Régions, l'état et la société locale* (Paris: PUF).

Ferrao, J. (1987), 'Social structures, labour markets and spatial configurations in modern Portugal', *Antipode*, 19: 99–118.

Gallagher, T. (1986), 'The Portuguese Communist Party', in B. Szajkowski (ed.), *Marxist Local Governments in W. Europe and Japan* (London: Frances Pinter).

Garofoli, G. (1992), 'Industrial districts: Structure and transformation', in id. (ed.), *Endogenous Development and Southern Europe* (Aldershot: Avebury).

Gramsci, A. (1975), *Selections from Political Writings, 1910–1920* (New York: International Publications).

Hadjimichalis, C. (1987), *Uneven Development and Regionalism: State, Territory and Class in Southern Europe* (London: Croom Helm).

—— and Papamichos, N. (1990), 'Local development in southern Europe: Towards a new mythology', *Antipode*, 22: 181–210.

—— and Vaiou, D. (1980), 'Penetration of multinational capital into backward regions: A policy analysis in Greece', *Mediterranean Peoples*, 10: 33–49.

—— —— (1990), 'Whose flexibility? The politics of informalisation in southern Europe', *Capital and Class*, 42: 79–106.

Haupt, G., and Weill, M. (1974), *Les Marxistes et la question nationale* (Paris: Maspéro).

Hudson, R., and Sadler, D. (1986), 'Contesting works closures in Western Europe's old industrial regions: Defending place or betraying class?', in A. Scott and M. Storper (eds.), *Production, Work, Territory* (London: Allen and Unwin).

Lankowski, C. (1987), 'Political opposition in the EC: The role of the Community from the Rome Treaty to the Danish Referendum of 1986', *Il politico*, 52/2: 251–93.

Lipietz, A. (1985), 'Le National et le régional: Quelle autonomie face à la crise capitaliste mondiale?', *Lesvos International Seminar Proceedings* (Dept. of Planning, University of Thessaloniki).

Massey, D. (1991), 'A global sense of place', *Marxism Today* (June), 24–9.

Nardin, G. (1987), *La Benetton: Strategia e struttura di un impresa di sucesso* (Rome: Lavoro).

Paci, M. (1982), *La struttura sociale italiana* (Bologna: Il Mulino).

Petras, J. (1984), 'The rise and decline of southern European socialist parties', *New Left Review*, 146: 37–52.

Pickvance, C. G. (1986), 'The rise and fall of urban social movements and the role of comparative analysis', *Society and Space*, 3: 31–53.

—— (forthcoming), 'Where have urban movements gone?', in C. Hadjimichalis and D. Sadler (eds.), *In, On and Around the Margins of a New Europe* (London: Belhaven).

Poulantzas, N. (1978), *State, Power and Socialism* (London: Verso).

Pred, A. (1989), 'The locally spoken word and local struggles', *Society and Space*, 7/2: 211–33.

Rokkan, S., and Urwin, D. W. (1983), *Economy, Territory and Identity* (London: Sage).

Sabel, C. F. (1989), 'Flexible specialisation and the re-emergence of regional economies', in P. Hirst and J. Zeitlin (eds.), *Reversing Industrial Decline?* (Oxford: Berg).

Sayer, A. (1992), 'Radical geography and marxist political economy: Towards a re-evaluation', *Progress in Human Geography*, 16/3: 343–60.

Soja, E. (1989), *Postmodern Geographies* (London: Verso).

Thrift, N. (1983), 'On the determination of social action in space and time', *Society and Space*, 1: 23–54.

Titos Moreno, A., and Rodriguez, J. J. (1979), *Crisis economica y empleo en Andalusia* (Madrid: Ministerio de Agricultura).

Touraine, A. (1981), *The Voice and the Eye: An Analysis of Social Movements* (Cambridge: Cambridge University Press).

Vasquez-Barquero, A. (1992), 'Local development initiatives under incipient regional autonomy: The Spanish experience in the 1980s', in G. Garofoli (ed.), *Endogenous Development and Southern Europe*.

Zangheri, R. (1975), 'Red Bologna: Example or exception?', interview in M. Jaggi and E. Ros, *Red Bologna* (London: Pluto).

12

Holding Down the Global

Ash Amin and Nigel Thrift

This book's chief concern has been with the territorial dimensions of globalization. Most particularly, its theme has been the apparent reassertion of the importance of territory that has run in parallel with globalization. Yet, what is clear is that in the several literatures concerned with this theme there is considerable confusion over what is the precise role of territory in globalization. This book has offered one solution to this dilemma through identifying the role of territory with the process of institutionalization, that is with the way in which local initiatives structure responses to processes of globalization and themselves become a part of processes of globalization. The chapters in this book have identified a number of different aspects of this thesis through the consideration of examples as diverse as the north-east of England, central Italy, eastern Germany, and the Swiss Jura.

But, as with any such volume, the questions that are answered throw up further questions. In this end piece we simply want to identify some of these questions as a part of an ongoing reformulation of the implications for globalization of territoriality.

One of the main unresolved problems that arises out of this book is the status of the term 'globalization' itself. In the chapters in the book it is interpreted in two distinct but overlapping ways. On the one hand, it is interpreted as a structural process with a logic of its own which is unfolding and which is more or less effective in certain places: there *is* a process of globalization and it favours certain places over others. On the other hand, it is interpreted as socially contested and the contestation is itself a part of the process of globalization. Quite clearly, these two interpretations provide a radically different outlook. The second interpretation is clearly more open-ended than the first. More importantly, the role of places is different. In the first interpretation places can only either institute a determined rearguard action against the process of globalization or look for a niche within it. What they cannot do is change it. They are structurally disempowered. In the second interpretation, places are presented with more options because they are a genuine part of the action. The outlook for many places may still be malign, but they can at least entertain the possibility of exerting influence on what shape the process of globalization takes through the form of local development strategy that they are able to institute.

Another problem with the status of the term 'globalization' is that although it is clear that some places will be empowered and others disempowered, in whatever ways the term is interpreted, it is not clear exactly how this differential can be gauged, unless one believes in a crushing logic of industrial change. This is, of course, why the second major theme of this book—institutionalization—is so important.

Four closely related questions immediately present themselves. The first of these concerns institutional diversity. It is by no means clear that the most successful regions will be the ones which have a tight-knit industrial structure, consisting of only a few closely defined institutional forms structured around a single industrial agenda. Grabher (1993a, 1993b) has argued 'in praise of waste'. That is, he suggests that the most successful regions over the longer run will be those which have considerable redundancy in their institutional structure. To use a biological analogy they will have a large 'genetic pool' of institutions, with overlapping spheres of influence. At first glance, such regions may seem to contain many local inefficiencies but these inefficiencies are actually the key to longer-term success. The institutional variety that these regions can offer, combined with their lack of any hegemonic institutional form, means that they constitute a constantly changing portfolio of possible organizational solutions to the problems of collective response to changed economic circumstances. In other words, regional adaptability is not based on a competition of firms but on a competition of forms. But, significantly, Grabher's interpretation begs the question of how it is possible to identify, let alone institute, effective rather than ineffective redundancy.

The second question follows on naturally from the first. How is it possible to measure the effectiveness of institutions? Grabher's intervention suggests that simple efficiency-based models of institutional effectiveness are inadequate to the task. Indeed they may be positively misleading since many institutions may be highly efficient but also ineffective. Such institutions are what Grabher evocatively calls *leere-hosen*, 'hollow trousers'. In other words, these institutions exist in form but not in spirit. Their 'grip' on power is slippery.

This question leads straight to a third. Nearly all the work that has taken place on institutional thickness has concentrated on the particularly successful (or particularly unsuccessful) regions: industrial districts and high-technology zones, versus areas of deindustrialization. Feverishly attempting to imitate the successful concentrations has been one of the chief concerns of many policy-makers, yet it might be argued that most of those successful concentrations are 'one-offs': fortuitous combinations of path dependence and specific local instances have produced a unique local institutional thickness which can never be imitated. It might be argued that this emphasis on the few examples of outstanding success diverts attention from the less spectacularly successful but still solid performers like the south-east of England or Lombardy in Italy, elements of whose success may be more easily imitated by policy-makers. This viewpoint can be linked to Grabher's ecological ideas on variety. It may be that policy-makers would be better employed in encouraging a diversity of institutional forms—from companies to cultural conventions—in a particular area, thus making for a more adaptive region. This is the politics of feasible imitation.

This point relates to the final question which concerns the nation state. Localities do not just need to position themselves in a globalizing world economy but also in a system of nation states (and increasingly metastates, like the European Community) which bargain directly with particular industrial forms like multinational corporations. Nation states still hold considerable power and they cannot be neglected in any solution to a region's ills. What national policy-makers can offer local policy-makers is resources to develop a pool of generic assets like skills and infrastructure, so that at least the basis for a diversity of institutional solutions is present.

These four questions point in the direction of a more general concern with institutions. That is the use of the term 'embeddedness'. Too often it seems that the economic is still seen as a separate sphere which is then, in some sense, institutional, collectivized, embedded in the social. The authors of the chapters in this book have shown a degree of unease with this metaphor for two reasons.

First, it is clear that the economy has to be considered as a set of institutions and institutionalizing processes (from norms and rules to organizations and other collectivities). But more than this, it is clear that all aspects of the process of institutionalization need to be simultaneously investigated, and not just those that are conventionally labelled as 'economic'. Without regard to this wider definition, reference in any future work on local economic development to the metaphor of embeddedness will be in danger of reproducing the split between the economic and the social that its usage was intended to avoid.

Second, the metaphor may be used to imply that some regions are more locally embedded that others. But again, this is a difficult concept to sustain. As we noted in Chapter 1, the meaning of the term 'local' is changing precisely as a result of globalization. Thus, in this book the term local embeddedness has only been used to highlight the importance of a set of local institutions and attributes in capturing global opportunities with significant positive effects on the local economy, so as to avoid notions of bounded and internally integrated territories.

In turn, this appeal to a wider notion of the local has other resonances. In particular, it seems clear that the cultural identities of particular places are a vital part of the process of local embeddedness but these identities are themselves in a state of flux because economic, social, and cultural connections are more and more likely to transgress a particular area. The danger is that exactly as globalization is occurring, local identities will become bogged down in romantic, reversionary, and even reactionary demands for the celebration of the uniqueness of a 'local' community. In other words, what is needed is to develop a more open and positive cultural regionalism which can cope with the increase in the economic, social, and cultural connections between places that globalization has helped to bring about. It will be a regionalism based on more open, mobile, and inclusive senses of identity, able to be positioned in a set of economic, social, and cultural networks of global extent, but not swamped by them (Massey 1993).

In conclusion, it is important to remember that the salience of institutions to local economies is a two-edged sword. The problem is not just how to achieve success—a question that the bulk of the literature addresses—but also how to sustain it. Regions like Emilia-Romagna and Baden Württemberg have managed to achieve success over thirty years but they now face the problem of maintaining this record without rigidly formalizing the social and cultural ingredients which have been one of the key determinants of that success—the conventions, the modes of social intervention, and the regional identities that give these places a 'character' which, fortuitously or not, has proved to be a potent economic weapon. As the example of the Swiss Jura shows, formalization can all too easily become ossification. The formidable records that these successful regions have built up do not necessarily guarantee continued success. Other more open regions may be able to replace them if they can find the 'right' institutional mix. What is quite clear is that the problem of institutionalization cannot be put to one side since it is only through the construction of adaptable institutional mixes that places can hold down the global.

REFERENCES

Grabher, G. (1993*a*) (ed.), *The Embedded Firm: On the Socio-economics of Industrial Networks* (London: Routledge).
—— (1993*b*), *In Praise of Waste: Redundancy in Regional Development* (Berlin: Edition Sigma).
Massey, D. (1993), 'Power-geometry and a progressive sense of place', in J. Bird *et al.* (eds.), *Mapping the Futures: Local Cultures, Global Change* (London: Routledge).

Index

Contributions to this volume are shown in **bold type**.
Abbreviations: the following may be used in subheadings—EC = European Community;
R & D = research and development; SMEs = small and medium enterprises.
Italic figures indicate tables.

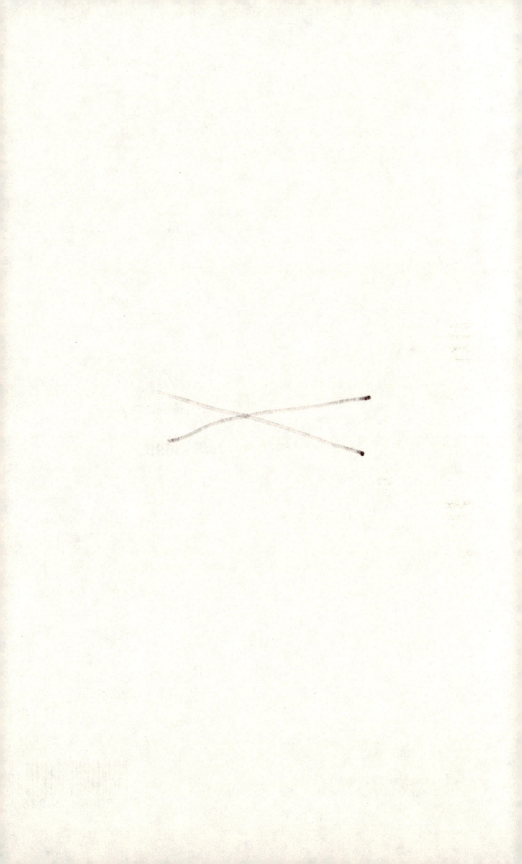

Printed in the United States
54043LVS00003B/135